Seaside Studies in Natural History

by

Elizabeth Cabot Cary Agassiz

THIS LITTLE BOOK IS AFFECTIONATELY DEDICATED BY THE AUTHORS TO PROFESSOR L. AGASSIZ, WHOSE PRINCIPLES OF CLASSIFICATION HAVE BEEN THE MAIN GUIDE IN ITS PREPARATION.

PREFACE.

This volume is published with the hope of supplying a want often expressed for some seaside book of a popular character, describing the marine animals common to our shores. There are many English books of this kind; but they relate chiefly to the animals of Great Britain, and can only have a general bearing on those of our own coast, which are for the most part specifically different from their European relatives. While keeping this object in view, an attempt has also been made to present the facts in such a connection, with reference to principles of science and to classification, as will give it in some sort the character of a manual of Natural History, in the hope of making it useful not only to the general reader, but also to teachers and to persons desirous of obtaining a more intimate knowledge of the subjects discussed in it. With this purpose, although nearly all the illustrations are taken from among the most common inhabitants of our bay, a few have been added from other localities in order to fill out this little sketch of Radiates, and render it, as far as is possible within such limits, a complete picture of the type.

A few words of explanation are necessary with reference to the joint authorship of the book. The drawings and the investigations, where they are not referred to other observers, have been made by MR. A. AGASSIZ, the illustrations having been taken, with very few exceptions, from nature, in order to represent the animals, as far as possible, in their natural attitudes; and the text has been written by MRS. L. AGASSIZ, with the assistance of MR. A. AGASSIZ's notes and explanations.

CAMBRIDGE, May, 1865.

* * * * *

NOTE.

This second edition is a mere reprint of the first. A few mistakes accidentally overlooked have been corrected; an explanation of the abbreviations of the names of writers used after the scientific names has been added, as well as a list of the wood-cuts. The changes which have taken place in the opinions of scientific men with regard to the distribution of animal life in the ocean have been duly noticed in their appropriate place, but no attempt has been made to incorporate more important additions which the progress of our knowledge of Radiates may require hereafter.

CAMBRIDGE, January, 1871.

* * * * *

CONTENTS.

DISTRIBUTION OF LIFE IN THE OCEAN

SYSTEMATIC TABLE

ABBREVIATIONS OF THE NAMES OF AUTHORS.

AG. L. Agassiz.

A. AG. A. Agassiz.

AYRES W. O. Ayres.

BLAINV. Blainville.

BOSC Bosc.

BR. Brandt.

CLARK H. J. Clark.

CUV. Cuvier.

D. & K. Dueben and Koren.

EDW. Milne-Edwards.

FORBES Edw. Forbes.

GRAY J. E. Gray.

JAEG. Jaeger.

LAM. Lamarck.

LAMX. Lamouroux.

LIN. Linnaeus.

LYM. Lyman.

M. & T. Mueller and Troschel.

MILL. Miller.

PER. et LES. Peron and Lesueur.

SARS M. Sars.

STIMP. Stimpson.

TIL. Tilesius.

MARINE ANIMALS OF MASSACHUSETTS BAY.

ON RADIATES IN GENERAL.

It is perhaps not strange that the Radiates, a type of animals whose home is in the sea, many of whom are so diminutive in size, and so light and evanescent in substance, that they are hardly to be distinguished from the element in which they live, should have been among the last to attract the attention of naturalists. Neither is it surprising to those who know something of the history of these animals, that when the investigation of their structure was once begun, when some insight was gained into their complex life, their association in fixed or floating communities, their wonderful processes of development uniting the most dissimilar individuals in one and the same cycle of growth, their study should have become one of the most fascinating pursuits of modern science, and have engaged the attention of some of the most original investigators during the last half century. It is true that from the earliest days of Natural History, the more conspicuous and easily accessible of these animals attracted notice and found their way into the scientific works of the time. Even Aristotle describes some of them under the names of Acalephae and Knidae, and later observers have added something, here and there, to our knowledge on the subject; but it is only within the last fifty years that their complicated history has been unravelled, and the facts concerning them presented in their true connection.

Among the earlier writers on this subject we are most indebted to Rondelet, in the sixteenth century, who includes some account of the Radiates, in his work on the marine animals of the Mediterranean. His position as Professor in the University at Montpelier gave him an admirable opportunity, of which he availed himself to the utmost, for carrying out his investigations in this direction. Seba and Klein, two naturalists in the North of Europe, also published at about this time numerous illustrations of marine animals, including Radiates. But in all these works we find only drawings and descriptions of the animals, without any attempt to classify them according to common structural features. In 1776, O. F. Mueller, in a work on the marine and terrestrial faunae of Denmark, gave some admirable figures of Radiates, several of which are identical with those found on our own coast. Cavolini also in his investigations on the lower marine animals of the Mediterranean, and Ellis in his work upon those of the British coast, did much during the latter half of the past century to enlarge our knowledge of them.

It was Cuvier, however, who first gave coherence and precision to all previous investigations upon this subject, by showing that these animals are united on a common plan of structure expressively designated by him under the name Radiata. Although, from a mistaken appreciation of their affinities, he associated some animals with them which do not belong to the type, and have since, upon a more intimate knowledge of their structure, been removed to their true positions; yet the principle introduced by him into their classification, as well as into that of the other types of the animal kingdom, has been all important to science.

It was in the early part of this century that the French began to associate scientific objects with their government expeditions. Scarcely any important voyage was undertaken to foreign countries by the French navy which did not include its corps of naturalists, under the patronage of government. Among the most beautiful figures we have of Radiates, are those made by Savigny, one of the French naturalists who accompanied Napoleon to Egypt; and from this time the lower marine animals began to be extensively collected and studied in their living condition. Henceforth the number of investigators in the field became more numerous, and it may not be amiss to give here a slight account of the more prominent among them.

Darwin's fascinating book, published after his voyage to the Pacific, and giving an account of the Coral islands, the many memoirs of Milne Edwards and Haime, and the great works of Quoy and Gaimard, and of Dana, are the chief authorities upon Polyps. In the study of the European Acalephs we have a long list of names high in the annals of science. Eschscholtz, Peron and Lesueur, Quoy and Gaimard, Lesson, Mertens, and Huxley, have all added largely to our information respecting these animals, their various voyages having enabled them to extend their investigations over a wide field. No less valuable have been the memoirs of Koelliker, Leuckart, Gegenbaur, Vogt, and Haeckel, who in their frequent excursions to the coasts of Italy and France have made a special study of the Acalephs, and whose descriptions have all the vividness and freshness which nothing but familiarity with the living specimens can give. Besides these, we have the admirable works of Von Siebold, of Ehrenberg, the great interpreter of the microscopic world, of Steenstrup, Dujardin, Dalyell, Forbes, Allman, and Sars. Of these, the four latter were fortunate in having their home on the sea-shore within reach of the objects of their study, so that they could watch them in their living condition, and follow all their changes. The charming books of Forbes, who knew so well how to popularize his instructions, and present scientific results under the most attractive form, are well known to English readers. But a word on the investigations of Sars may not be superfluous.

Born near the coast of Norway, and in early life associated with the Church, his passion for Natural History led him to employ all his spare time in the study of the marine animals immediately about him, and his first papers on this subject attracted so much attention, that he was offered the place of Professor at Christiania, and henceforth devoted himself exclusively to scientific pursuits, and especially to the investigation of the Acalephs. He gave us the key to the almost fabulous transformations of these animals, and opened a new path in science by showing the singular phenomenon of the so-called "alternate generations," in which the different phases of the same life may be so distinct and seemingly so disconnected that, until we find the relation between them, we seem to have several animals where we have but one.

To the works above mentioned, we may add the third and fourth volumes of Professor Agassiz's Contributions to the Natural History of the United States, which are entirely devoted to the American Acalephs.

The most important works and memoirs concerning the Echinoderms are those by Klein, Link, Johannes Mueller, Jaeger, Desmoulins, Troschel, Sars, Savigny, Forbes, Agassiz, and Luetken, but excepting those of Forbes and Sars, few of these observations are made upon the living specimens. It may be well to mention here, for the benefit of those who care to know something more of the literature of this subject in our own country, a number of memoirs on the Radiates of our coasts, published by the various scientific societies of the United States, and to be found in their annals. Such are the papers of Gould, Agassiz, Leidy, Stimpson, Ayres, McCrady, Clark, A. Agassiz, and Verrill.

One additional word as to the manner in which the subjects included in the following descriptions are arranged. We have seen that Cuvier recognized the unity of plan in the structure of the whole type of Radiates. All these animals have their parts disposed around a common central axis, and diverging from it toward the periphery. The idea of bilateral symmetry, or the arrangement of parts on either side of a longitudinal axis, on which all the higher animals are built, does not enter into their structure, except in a very subordinate manner, hardly to be perceived by any but the professional naturalist. This radiate structure being then common to the whole type, the animals composing it appear under three distinct structural expressions of the general plan, and according to these differences are divided into three classes,-- Polyps, Acalephs, and Echinoderms. With these few preliminary remarks we may now take up in turn these different groups, beginning with the lowest, or the Polyps.[1]

[Footnote 1: It is to be regretted that on account of the meagre representations of Polyps on our coast, where the coral reefs, which include the most interesting features of Polyp life, are entirely wanting, our account of these animals is necessarily deficient in variety of material. When we reach the Acalephs or Jelly-Fishes, in which the fauna of our shores is especially rich, we shall not have the same apology for dulness; and it will be our own fault if our readers are not attracted by the many graceful forms to which we shall then introduce them.]

* * * * *

GENERAL SKETCH OF THE POLYPS.

Before describing the different kinds of Polyps living on our immediate coast, we will say a few words of Polyps in general and of the mode in which the structural plan common to all Radiates is adapted to this particular class. In all Polyps the body consists of a sac divided by vertical partitions (Fig. 1.) into distinct cavities or chambers. These partitions are not, however, all formed at once, but are usually limited to six at first, multiplying indefinitely with the growth of the animal in some kinds, while in others they never increase beyond a certain definite number. In the axis of the sac, thus divided, hangs a smaller one, forming the digestive cavity, and supported for its whole length by the six primary partitions. The other partitions, though they extend more or less inward in proportion to their age, do not unite with the digestive sac, but leave a free space in the centre between their inner edge and the outer wall of the digestive sac. The genital organs are placed on the inner edges of the partitions, thus hanging as it were at the door of the chambers, so that when hatched, the eggs naturally drop into the main cavity of the body, whence they pass into the second smaller sac through an opening in its bottom or digestive cavity, and thence out through the mouth into the water. In the lower Polyps, as in our common Actinia for instance, these organs occur on all the radiating partitions, while among the higher ones, the Halcyonoids for example, they are found only on a limited number. This limitation in the repetition of identical parts is always found to be connected with structural superiority.

The upper margin of the body is fringed by hollow tentacles, each of which opens into one of the chambers. All parts of the animal thus communicate with each other, whatever is introduced at the mouth circulating through the whole structure, passing first into the digestive cavity, thence through the opening in the bottom into the main chambered cavity, where it enters freely into all the chambers, and from the chambers into the tentacles. The rejected portions of the food, after the process of digestion is completed, return by the same road and are thrown out at the mouth.

These general features exist in all Polyps, and whether they lead an independent life as the Actinia, or are combined in communities, like most of the corals and the Halcyonoids; whether the tentacles are many or few; whether the partitions extend to a greater or less height in the body; whether they contain limestone deposit, as in the corals, or remain soft throughout life as the sea-anemone,--the above description applies to them all, while the minor differences, either in the tentacles or in the form, size, color, and texture of the body, are simply modifications of this structure, introducing an infinite variety into the class, and breaking it up into the lesser groups designated as orders, families, genera, and species. Let us now look at some of the divisions thus established.

The class of Polyps is divided into three orders,--the Halcyonoids, the Madreporians, and the Actinoids. Of the lowest among these orders, the Actinoid Polyps, our Actinia or sea-anemone is a good example. They remain soft through life, having a great number of partitions and consequently a great number of tentacles, since there is a tentacle corresponding to every chamber. Indeed, in this order the multiplication of tentacles and partitions is indefinite, increasing during the whole life of the animal with its growth; while we shall see that in some of the higher orders the constancy and limitation in the number of these parts is an indication of superiority, being accompanied by a more marked individualization of the different functions.

Next come the Madreporians, of which our Astrangia, to be described hereafter, may be cited as an example. In this group, although the number of tentacles still continues to be large, they are nevertheless more limited than in the Actinoids; but their characteristic feature is the deposition of limestone walls in the centre of the chambers formed by the soft partitions, so that all the soft partitions alternate with hard ones. The tentacles, always corresponding to the cavity of the chambers, may be therefore said to ride this second set of partitions arising just in the centre of the chambers.

The third and highest order of Polyps is that of the Halcyonoids. Here the partitions are reduced to eight; the tentacles, according to the invariable rule, agree in number with the chambers, but have a far more highly complicated structure than in the lower Polyps. Some of these Halcyonoids deposit limestone particles in their frame. But the tendency to solidify is not limited to definite points, as in the Madreporians. It may take place anywhere, the rigidity of the whole structure increasing of course in proportion to the accumulation of limestone. There are many kinds, in which the axis always remains soft or cartilaginous, while others, as the so-called sea-fans for instance, well known among the corals for their beauty of form and color, are stiff and hard throughout. Whatever their character in this respect, however, they are always compound, living in communities, and never found as separate individuals after their early stages of growth. Some of those with soft axis lead a wandering life, enjoying as much freedom of movement as if they had an individual existence, shooting through the water like the Pennatulae, well known on the California coast, or working their way through the sand like the Renilla, common on the sandy shores of our Southern States.

* * * * *

ACTINOIDS.

Actinia, or Sea-Anemone. (Metridium marginatum EDW.)

Nothing can be more unprepossessing than a sea-anemone when contracted. A mere lump of brown or whitish jelly, it lies like a lifeless thing on the rock to which it clings, and it is difficult to believe that it has an elaborate and exceedingly delicate internal organization, or will ever expand into such grace and beauty as really to deserve the name of the flower after which it has been called. Figs. 2, 3, 4, and 5, show this animal in its various stages of expansion and contraction. Fig. 2 represents it with all its external appendages folded in, and the whole body flattened; in Fig. 3, the tentacles begin to steal out, and the body rises slightly; in Fig. 4, the body has nearly gained its full height, and the tentacles, though by no means fully spread, yet form a delicate wreath around the mouth; while in Fig. 5, drawn in life size, the whole summit of the body seems crowned with soft, plumy fringes. We would say for the benefit of collectors that these animals are by no means difficult to find, and thrive well in confinement, though it will not do to keep them in a small aquarium with other specimens, because they soon render the water foul and unfit for their companions. They should therefore be kept in a separate glass jar or bowl, and under such circumstances will live for a long time with comparatively little care.

They may be found in any small pools about the rocks which are flooded by the tide at high water. Their favorite haunts, however, where they occur in greatest quantity are more difficult to reach; but the curious in such matters will be well rewarded, even at the risk of wet feet and a slippery scramble over rocks covered with damp sea-weed, by a glimpse into their more crowded abodes. Such a grotto is to be found on the rocks of East Point at Nahant. It can only be reached at low tide, and then one is obliged to creep on hands and knees to its entrance, in order to see through its entire length; but its whole interior is studded with these animals, and as they are of various hues, pink, brown, orange, purple, or pure white, the effect is like that of brightly colored mosaics set in the roof and walls. When the sun strikes through from the opposite extremity of this grotto, which is open at both ends, lighting up its living mosaic work, and showing the play of the soft fringes wherever the animals are open, it would be difficult to find any artificial grotto to compare with it in beauty. There is another of the same kind on Saunders's Ledge, formed by a large boulder resting on two rocky ledges, leaving a little cave beneath, lined in the same way with variously colored sea-anemones, so closely studded over its walls that the surface of the rock is completely hidden. They are, however, to be found in larger or smaller clusters, or scattered singly in any rocky fissures, overhung by sea-weed, and accessible to the tide at high water.

The description of Polyp structure given above includes all the general features of the sea-anemone; but for the better explanation of the figures, it may not be amiss to recapitulate them here in their special application. The body of the sea-anemone may be described as a circular, gelatinous bag, the bottom of which is flat and slightly spreading around the margin. (Fig. 2.) The upper edge of this bag turns in so as to form a sac within a sac. (Fig. 6.) This inner sac, s, is the stomach or digestive cavity, forming a simple open space in the centre of the body, with an aperture in the bottom, b, through which the food passes into the larger sac, in which it is enclosed. But this outer and larger sac or main cavity of the body is not, like the inner one, a simple open space. It is, on the contrary, divided by vertical partitions into a number of distinct chambers, converging from the periphery to the centre. These partitions do not all advance so far as actually to join the wall of the digestive cavity hanging in the centre of the body, but most of them stop a little short of it, leaving thus a small, open space between the chambers and the inner sac. (Fig. 1.) The eggs hang on the inner edge of the partitions; when mature they drop into the main cavity, enter the inner digestive cavity through its lower opening, and are passed out through the mouth.

The embryo bears no resemblance to the mature animal. It is a little planula, semi-transparent, oblong, entirely covered with vibratile cilia, by means of which it swims freely about in the water till it establishes itself on some rocky surface, the end by which it becomes attached spreading slightly and fitting itself to the inequalities of the rock so as to form a secure basis. The upper end then becomes depressed toward the centre, that depression deepening more and more till it forms the inner sac, or in other words the digestive cavity described above. The open mouth of this inner sac, which may, however, be closed at will, since the whole substance of the body is exceedingly contractile, is the oral opening or so-called mouth of the animal. We have seen how the main cavity becomes divided by radiating partitions into numerous chambers; but while these internal changes are going on, corresponding external appendages are forming in the shape of the tentacles, which add so much to the beauty of the animal, and play so important a part in its history. The tentacles, at first only few in number, are in fact so many extensions of the inner chambers, gradually narrowing upward till they form these delicate hollow feelers which make a soft downy fringe all around the mouth. (Fig. 7.) They do not start abruptly from the summit, but the upper margin of the body itself thins out to form more or less extensive lobes, through which the partitions and chambers continue their course, and along the edge of which the tentacles arise.

The eggs are not always laid in the condition of the simple planula described above. They may, on the contrary, be dropped from the parent in different stages of development, sometimes even after the tentacles have begun to form, as in Figs. 8, 9. Neither is it by means of eggs alone that these animals reproduce themselves; they may also multiply by a process of self-division. The disk of an Actinia may contract along its centre till the circular outline is changed to that of a figure 8, this constriction deepening gradually till the two halves of the 8 separate, and we have an Actinia with two mouths, each surrounded by an independent set of tentacles. Presently this separation descends vertically till the body is finally divided from summit to base, and we have two Actiniae where there was originally but one. Another and a far more common mode of reproduction among these animals is that of budding like corals. A slight swelling arises on the side of the body or at its base; it enlarges gradually, a digestive cavity is formed within it, tentacles arise around its summit, and it finally drops off from the parent and leads an independent existence. As a number of these buds are frequently formed at once, such an Actinia, surrounded by its little family, still attached to the parent, may appear for a time like a compound stock, though their normal mode of existence is individual and distinct.

The Actinia is exceedingly sensitive, contracting the body and drawing in the tentacles almost instantaneously at the slightest touch. These sudden movements are produced by two powerful sets of muscles, running at right angles with each other through the thickness of the body wall; the one straight and vertical, extending from the base of the wall to its summit; the other circular and horizontal, stretching concentrically around it. By the contraction of the former, the body is of course shortened; by the contraction of the latter, the body is, on the contrary, lengthened in proportion to the compression of its circumference. Both sets can easily be traced by the vertical and horizontal lines crossing each other on the external wall of the body, as in Fig. 5. Each tentacle is in like manner furnished with a double set of muscles, having an action similar to that described above. In consequence of these violent muscular contractions, the water imbibed by the animal, and by which all its parts are distended to the utmost, is forced, not only out of the mouth, but also through small openings in the body wall scarcely perceptible under ordinary circumstances, but at such times emitting little fountains in every direction.

Notwithstanding its extraordinary sensitiveness, the organs of the senses in the Actinia are very inferior, consisting only of a few pigment cells accumulated at the base of the tentacles. The two sets of muscles meet at the base of the body, forming a disk, or kind of foot, by which the animal can fix itself so firmly to the ground, that it is very difficult to remove it without injury. It is nevertheless capable of a very limited degree of motion, by means of the expansion and contraction of this foot-like disk.

The Actiniae are extremely voracious; they feed on mussels and cockles, sucking the animals out of their shells. When in confinement they may be fed on raw meat, and seem to relish it; but if compelled to do so, they will live on more meagre fare, and will even thrive for a long time on such food as they may pick up in the water where they are kept.

Rhodactinia (Rhodactinia Davisii AG.)

Very different from this is the bright red Rhodactinia (Fig. 10), quite common in the deeper waters of our bay, while farther north, in Maine, it occurs at low-water mark. Occasionally it may be found thrown up on our sandy beaches after a storm, and then, if it has not been too long out of its native element, or too severely buffeted by the waves, it will revive on being thrown into a bucket of fresh sea-water, expand to its full size, and show all the beauty of its natural coloring. It is crowned with a wreath of thick, short tentacles (Fig. 10), and though so vivid and bright in color, it is not so pretty as the more common Actinia marginata, with its soft waving wreath of plume-like feelers, in comparison to which the tentacles of the Rhodactinia are clumsy and slow in their movements.

All Actiniae are not attached to the soil like those described above, nor do they all terminate in a muscular foot, some being pointed or rounded at their extremity. Many are nomadic, wandering about at will during their whole lifetime, others live buried in the sand or mud, only extending their tentacles beyond the limits of the hole where they make their home; while others again lead a parasitic life, fastening themselves upon our larger jelly-fish, the Cyaneae, though one is at a loss to imagine what sustenance they can derive from animals having so little solidity, and consisting so largely of water.

Arachnactis. (Arachnactis brachiolata A. AG.)

Among the nomadic Polyps is a small floating Actinia, called Arachnactis, (Fig. 11,) from its resemblance to a spider. They are found in great plenty floating about during the night, feeling their way in every direction by means of their tentacles, which are large in proportion to the size of the animal, few in number, and turned downward when in their natural attitude. The partitions and the digestive cavity enclosed between them are short, as will be seen in Fig. 11, when compared to the general cavity of the body floating balloon-like above them. Around the mouth is a second row of shorter tentacles, better seen in a younger specimen (Fig. 12). This Actinia differs from those described above, in having two of the sides flattened, instead of being perfectly circular. Looked at from above (as in Fig. 13) this difference in the diameters is very perceptible; there is an evident tendency towards establishing a longitudinal axis. In the sea-anemone, this disposition is only hinted at in the slightly pointed folds or projections on opposite sides of the circle formed by the mouth, which in the Arachnactis are so elongated as to produce a somewhat narrow slit (see Fig. 13), instead of a circular opening. The mouth is also a little out of centre, rather nearer one end of the disk than the other. These facts are interesting, as showing that the tendency towards establishing a balance of parts, as between an anterior and posterior extremity, a right and left side, is not forgotten in these lower animals, though their organization as a whole is based upon an equality of parts, admitting neither of posterior and anterior extremities, nor of right and left, nor of above and below, in a structural sense. This animal also presents a seeming anomaly in the mode of formation of the young tentacles, which always make their appearance at the posterior extremity of the longitudinal axis, the new ones being placed behind the older ones, instead of alternating with them as in other Actiniae.

Bicidium. (Bicidium parasiticum AG.)

The Bicidium (Fig. 14), our parasitic Actinia, is to be sought for in the mouth-folds of the Cyanea, our common large red Jelly-fish. In any moderate-sized specimen of the latter from twelve to eighteen inches in diameter, we shall be sure to find one or more of these parasites, hidden away among the numerous folds of the mouth. The body is long and tapering, having an aperture in the extremity, the whole animal being like an elongated cone, strongly ribbed from apex to base. At the base, viz. at the month end, are a few short, stout tentacles. This Actinia is covered with innumerable little transverse wrinkles (see Fig. 14), by means of which it fastens itself securely among the fluted membranes around the mouth of the Jelly-fish. It will live a considerable time in confinement, attaching itself, for its whole length, to the vessel in which it is kept, and clinging quite firmly if any attempt is made to remove it. The general color of the body is violet or a brownish red, though the wrinkles give it a somewhat mottled appearance. Halcampa. (Halcampa albida AG.)

Strange to say, the Actiniae, which live in the mud, are among the most beautifully colored of these animals. They frequently prepare their home with some care, lining their hole by means of the same secretions which give their slimy surface to our common Actiniae, and thus forming a sort of tube, into which they retire when alarmed. But if undisturbed, they may be seen at the open door of their house with their many colored disk and mottled tentacles extending beyond the aperture, and their mouth wide open, waiting for what the tide may bring them. By the play of their tentacles, they can always produce a current of water about the mouth, by means of which food passes into the stomach. We have said, that these animals are very brightly colored, but the little Halcampa (Fig. 15), belonging to our coast, is not one of the brilliant ones. It is, on the contrary, a small, insignificant Actinia, resembling a worm, as it burrows its way through the sand. It is of a pale yellowish color, with whitish warts on the surface.

* * * * *

MADREPORIANS.

Astrangia. (Astrangia Danae AG.)

In Figure 16, we have the only species of coral growing so far north as our latitude. Indeed, it hardly belongs in this volume, since we have limited ourselves to the Radiates of Massachusetts Bay,--its northernmost boundary being somewhat to the south of Massachusetts Bay, about the shores of Long

Island, and on the islands of Martha's Vineyard Sound. But we introduce it here, though it is not included under our title, because any account of the Radiates, from which so important a group as that of the corals was excluded, would be very incomplete.

This pretty coral of our Northern waters is no reef-builder, and does not extend farther south than the shores of North Carolina. It usually establishes itself upon broken angular bits of rock, lying in sheltered creeks and inlets, where the violent action of the open sea is not felt. The presence of one of these little communities on a rock may first be detected by what seems like a delicate white film over the surface. This film is, however, broken up by a number of hard calcareous deposits in very regular form (Fig. 20), circular in outline, but divided by numerous partitions running from the outer wall to the centre of every such circle, where they unite at a little white spot formed by the mouth or oral opening. These circles represent, and indeed are themselves the distinct individuals (Fig. 17) composing the community, and they look not unlike the star-shaped pits on a coral head, formed by Astraeans. Unlike the massive compact kinds of coral, however, the individuals multiply by budding from the base chiefly, never rising one above the other, but spreading over the surface on which they have established themselves, a few additional individuals arising between the older ones. In consequence of this mode of growth, such a community, when it has attained any size, forms a little white mound on the rock, higher in the centre, where the older members have attained their whole height and solidity, and thinning out toward the margin, where the younger ones may be just beginning life, and hardly rise above the surface of the rock. These communities rarely grow to be more than two or three inches in diameter, and about quarter of an inch in height at the centre where the individuals have reached their maximum size. When the animals are fully expanded (Fig. 18), with all their tentacles spread, the surface of every such mound becomes covered with downy white fringes, and what seemed before a hard, calcareous mass upon the rock, changes to a soft fleecy tuft, waving gently to and fro in the water. The tentacles are thickly covered with small wart-like appendages, which, on examination, prove to be clusters of lasso-cells, the terminal cluster of the tentacle being quite prominent. These lasso-cells are very formidable weapons, judging both from their appearance when magnified (Fig. 19), and from the terrible effect of their bristling lash upon any small crustacean, or worm, that may be so unfortunate as to come within its reach.

The description of the internal arrangement of parts in the Actinia applies in every particular to these corals, with the exception of the hard deposit in the lower part of the body. As in all the Polyps, radiating partitions divide the main cavity of the body into distinct separate chambers, and the tentacles increasing by multiples of six, numbering six in the first set, six in the second, and twelve in the third, are hollow, and open into the chambers. But the feature which distinguishes them from the soft Actiniae, and unites them with the corals, requires a somewhat more accurate description. In each individual, a hard deposit is formed (Fig. 20), beginning at the base of every chamber, and rising from its floor to about one fifth the height of the animal at its greatest extension. This lime deposit does not, however, fill the chamber for its whole width, but rises as a thin wall in its centre. (See Figs. 13, 17.) Thus between all the soft partitions, in the middle of the chambers which separate them, low limestone walls are gradually built up, uniting in a solid column in the centre. These walls run parallel with the soft partitions, although they do not rise to the same height, and they form the radiating lines like stiff lamellae, so conspicuous when all the soft parts of the body are drawn in. The mouth of the Astrangia is oval, and the partitions spread in a fan-shaped way, being somewhat shorter at one side of the animal than on the other. The partitions extend beyond the solid wall which unites them at the periphery, in consequence of which, this wall is marked by faint vertical ribs.

* * * * *

HALCYONOIDS

Halcyonium. (Halcyonium carneum AG.)

We come now to the Halcyonoids, represented in our waters by the Halcyonium (Fig. 22). In the Halcyonoids, the highest group of Polyps, the tentacles reach their greatest limitation, which, as above mentioned, is found to be a mark of superiority, and, connected with other structural features, places them at the head of their class. The number of tentacles throughout this group is always eight. They are very complicated (Fig. 21), in comparison with the tentacles of the lower orders, being deeply lobed, and fringed around the margin. Our Halcyonium communities (Fig. 22) usually live in deep water, attached to dead shells, though they may occasionally be found growing at low-water mark, but this is very rare. They have received a rather lugubrious name from the fishermen, who call them "dead-men's fingers," and indeed, when the animals are contracted, such a community, with its short branches attached to the main stock, looks not unlike the stump of a hand, with short, fat fingers. In such a condition they are very ugly, the whole mass being somewhat gelatinous in texture, and a dull, yellowish pink in color. But when the animals, which are capable of great extension, are fully spread, as in Fig. 22, such a polyp-stock has a mossy, tufted look, and is by no means an unsightly object. When the individuals are entirely expanded, as in Fig. 23, they become quite transparent, and their internal structure can readily be seen through the walls of the body; we can then easily distinguish the digestive cavity, supported for its whole length by the eight radiating partitions, as well as the great size of the main digestive cavity surrounding it. Notwithstanding the remarkable power of contraction and dilatation in the animals themselves, the tentacles are but slightly contractile. This kind of community increases altogether by budding, the individual polyps remaining more or less united, the tissues of the individuals becoming thicker by the deposition of lime nodules, and thus forming a massive semi-cartilaginous pulp, uniting the whole community. In the neighborhood of Provincetown they are very plentiful, and are found all along the shores of our Bay in deep water.

GENERAL SKETCH OF ACALEPHS.

In the whole history of metamorphosis, that wonderful chapter in the life of animals, there is nothing more strange or more interesting than the transformations of the Acalephs. First, as little floating planulae or transparent spheres, covered with fine vibratile cilia, by means of which they move with great rapidity, then as communities fixed to the ground and increasing by budding like the corals, or multiplying by self-division, and later as free-swimming Jelly-fishes, many of them pass through phases which have long baffled the investigations of naturalists, and have only recently been understood in their true connection. Great progress has, however, been made during this century in our knowledge of this class. Thanks to the investigations of Sars, Dujardin, Steenstrup, Van Beneden, and many others, we now have the key to their true relations, and transient phases of growth, long believed to be the adult condition of distinct animals, are recognized as parts in a cycle of development belonging to one and the same life. As the class now stands, it includes three orders, highest among which are the CTENOPHORAE, so called on account of their locomotive organs, consisting of minute flappers arranged in vertical comb-like rows; next to these are the DISCOPHORAE, with their large gelatinous umbrella-like disks, commonly called Jelly-fishes, Sun-fishes, or Sea-blubbers, and below these come the HYDROIDS, embracing the most minute and most diversified of all these animals.

These orders are distinguished not only by their striking external differences, but by their mode of development also. The Ctenophorae grow from eggs by a direct continuous process of development, without undergoing any striking metamorphosis; the Discophorae, with some few exceptions, in which they develop like the Ctenophorae from eggs, begin life as a Hydra-like animal, the subsequent self-division of which gives rise, by a singular process, presently to be described, to a number of distinct Jelly-fishes; the Hydroids include all those Acalephs which either pass the earlier stages of their existence as little shrub-like communities, or remain in that condition through life. These Hydroid stocks, as they are sometimes called, give rise to buds; these buds are transformed into Jelly-fishes, which in some instances break off when mature and swim away as free animals, while in others they remain permanent members of the Hydroid stock, never assuming a free mode of life. All these buds when mature, whether free or fixed, lay eggs in their turn, from which a fresh stock arises to renew this singular cycle of growth, known among naturalists as "alternate generations."

The Hydroids are not all attached to the ground,--some like the Physalia (Portuguese man-of-war), or the Nanomia, that pretty floating Hydroid of our own waters, move about with as much freedom as if they enjoyed an individual independent existence. As all these orders have their representatives on our coast, to be described hereafter in detail, we need only allude here to their characteristic features. But we must not leave unnoticed one very remarkable Hydroid Acaleph (Fig. 24), not found in our waters, and resembling the Polyps so much, that it has long been associated with them. The Millepore is a coral, and was therefore the more easily confounded with the Polyps, so large a proportion of which build coral stocks; but a more minute investigation of its structure (Figs. 25, 26) has recently shown that it belongs with the Acalephs.[2] This discovery is the more important, not only as explaining the true position of this animal in the Animal Kingdom, but as proving also the presence of Acalephs in the earliest periods of creation, since it refers a large number of fossil corals, whose affinities with the millepores are well understood, to that class, instead of to the class of Polyps with which they had hitherto been associated. But for this we should have no positive evidence of the existence of Acalephs in early geological periods, the gelatinous texture of the ordinary Jelly-fishes making their preservation almost impossible. It is not strange that the true nature of this animal should have remained so long unexplained; for it is only by the soft parts of the body, not of course preserved in the fossil condition, that their relations to the Acalephs may be detected; and they are so shy of approach, drawing their tentacles and the upper part of the body into their limestone frame if disturbed, that it is not easy to examine the living animal.

[Footnote 2: See "Methods of Study," by Prof. Agassiz.]

The Millepore is very abundant on the Florida reefs. From the solid base of the coral stock arise broad ridges, branching more or less along the edges, the whole surface being covered by innumerable pores, from which the diminutive animals project when expanded. (Fig. 25.) The whole mass of the coral is porous, and the cavities occupied by the Hydrae are sunk perpendicularly to the surface within the stock. Seen in a transverse cut these tubular cavities are divided at intervals by horizontal partitions (Fig. 26), extending straight across the cavity from wall to wall, and closing it up entirely, the animal occupying only the outer-most open space, and building a new partition behind it as it rises in the process of growth. This structure is totally different from that of the Madrepores, Astraeans, Porites, and indeed, from all the polyp corals which, like all Polyps, have the vertical partitions running through the whole length of the body, and more or less open from top to bottom.

The life of the Jelly-fishes, with the exception of the Millepores and the like, is short in comparison to that of other Radiates. While Polyps live for many years, and Star-fishes and Sea-urchins require ten or fifteen years to attain their full size, the short existence of the Acaleph, with all its changes, is accomplished in one year. The breeding season being in the autumn, the egg grows into a Hydroid during the winter; in the spring the Jelly-fish is freed from the Hydroid stock, or developed upon it as the case may be; it attains its full size in the fall, lays its eggs and dies, and the cycle is complete. The autumn storms make fearful havoc among them, swarms of them being killed by the fall rains, after which they may be found thrown up on the beaches in great numbers. When we consider the size of these Jelly-fishes, their rapidity of growth seems very remarkable. Our common Aurelia measures some twelve to eighteen inches in diameter when full grown, and yet in the winter it is a Hydra so small as almost to escape notice. Still more striking is the rapid increase of our Cyanea, that giant among Jelly-fishes, which, were it not for the soft, gelatinous consistency of its body, would be one of the most formidable among our marine animals.

Before entering upon the descriptions of the special kinds of Jelly-fishes, we would remind our readers that the radiate plan of structure is reproduced in this class of animals as distinctly as in the Polyps, though under a different aspect. Here also we find that there is a central digestive cavity from which all the radiating cavities, whether simple or ramified, diverge toward the periphery. It is true that the open chambers of the Polyps are here transformed into narrow tubes, by the thickening of the dividing partitions; or in other words, the open spaces of the Polyps correspond to tubes in the Acalephs, while the partitions in the Polyps correspond to the thick masses of the body dividing the tubes in the Acalephs. But the principle of radiation on which the whole branch of Radiates is constructed controls the organization of Acalephs no less than that of the other classes, so that a transverse section across any Polyp (Fig. 1), or across any Acaleph (Fig. 50), or across any Echinoderm (Fig. 140), shows their internal structure to be based upon a radiation of all parts from the centre to the periphery.

That there may be no vagueness as to the terms used hereafter, we would add one word respecting the nomenclature of this class, whose aliases might baffle the sagacity of a police detective. The names Acalephs, Medusae, or the more common appellation of Jelly-fishes, cover the same ground, and are applied indiscriminately to the animals they represent. The name Jelly-fish is an inappropriate one, though the gelatinous consistency of these animals is accurately enough expressed by it; but they have no more structural relation to a fish than to a bird or an insect. They have, however, received this name before the structure of animals was understood, when all animals inhabiting the waters were indiscriminately called fishes, and it is now in such general use that it would be difficult to change it. The name Medusa is derived from their long tentacular appendages, sometimes wound up in a close coil, sometimes thrown out to a great distance, sometimes but half unfolded, and aptly enough compared to the snaky locks of Medusa. Their third and oldest appellation, that of Acalephs,--alluding to their stinging or nettling property, and given to them and like animals by Aristotle, in the first instance, but afterwards applied by Cuvier in a more limited sense to Jelly-fishes,--is the most generally accepted, and perhaps the most appropriate of all.

The subject of nomenclature is not altogether so dry and arid as it seems to many who do not fully understand the significance of scientific names. Not only do they often express with terse precision the character of the animal or plant they signify, but there is also no little sentiment concealed under these jaw-breaking appellations. As seafaring men call their vessels after friends or sweethearts, or commemorate in this way some impressive event, or some object of their reverence, so have naturalists, under their fabrication of appropriate names, veiled many a graceful allusion, either to the great leaders of our science, or to some more intimate personal affection. The Linnaea borealis was well named after his famous master, by a disciple of the great Norwegian naturalist; Goethea semperflorens, the ever-blooming, is another tribute of the same kind, while the pretty, graceful little Lizzia, named by Forbes, is one instance among many of a more affectionate reference to nearer friends. The allusions of this kind are not always of so amiable a character, however,--witness the "Buffonia," a low, noxious weed, growing in marshy places, and named by Linnaeus after Buffon, whom he bitterly hated. Indeed, there is a world of meaning hidden under our zoological and botanical nomenclature, known only to those who are intimately acquainted with the annals of scientific life in its social as well as its professional aspect.

* * * * *

CTENOPHORAE.

The Ctenophorae differ from other Jelly-fishes in their mode of locomotion. All the Discophorous Medusae, as well as Hydroids, move by a rhythmical rise and fall of the disk, contracting and expanding with alternations so regular, that it reminds one of the action of the lungs, and seems at first sight to be a kind of respiration in which water takes the place of air. The Greeks recognized this peculiar character in their name, for they called them Sea-lungs. Indeed, locomotion, respiration, and circulation are so intimately connected in all these lower animals, that whatever promotes one of these functions affects the other also, and though the immediate result of the contraction and expansion of the disk seems to be to impel them through the water, yet it is also connected with the introduction of water into the body, which there becomes assimilated with the food in the process of digestion, and is circulated throughout all its parts by means of ramifying tubes. In the Ctenophorae there is no such regular expansion and contraction of the disk; they are at once distinguished from the Discophorae by the presence of external locomotive appendages of a very peculiar character. They move by the rapid flapping of countless little oars or paddles, arranged in vertical rows along the surface of the disk, acting independently of each other; one row, or even one paddle, moving singly, or all of them together, at the will of the animal; thus enabling it to accelerate or slacken its movements, to dart through the water rapidly, or to diminish its speed by partly furling its little sails, or, spreading them slightly, to poise itself with a faint, quivering movement that reminds one of the pause of the humming-bird in the air,-- something that is neither positive motion, nor actual rest.[3]

[Footnote 3: The flappers of one side are sometimes in full activity, while those of the other side are perfectly quiet or nearly so, thus producing rotatory movements in every direction.]

These locomotive appendages are intimately connected with the circulating tubes, as we shall see when we examine the structural details of these animals, so that in them also breathing and moving are in direct relation to each other. To those unaccustomed to the comparison of functions in animals, the use of the word breathing, as applied to the introduction of water into the body, may seem inappropriate, but it is by the absorption of aerated water that these lower animals receive that amount of oxygen into the system, as necessary to the maintenance of life in them, as a greater supply is to the higher animals. The name of Ctenophorae or comb-bearers, is derived from these rows of tiny paddles which have been called combs by some naturalists, because they are set upon horizontal bands of muscles, see Fig. 29, reminding one of the base of a comb, while the fringes are compared to its teeth. These flappers add greatly to the beauty of these animals, for a variety of brilliant hues is produced along each row by the decomposition of the rays of light upon them when in motion. They give off all the prismatic colors, and as the combs are exceedingly small, so that at first sight one hardly distinguishes them from the disk itself, the exquisite play of color, rippling in regular lines over the surface of the animal, seems at first to have no external cause.

Pleurobrachia. (Pleurobrachia rhododactyla Ag.)

Among the most graceful and attractive of these animals are the Pleurobrachia (Fig. 29), and, though not first in order, we will give it the precedence in our description, because it will serve to illustrate some features of the other two groups. The body of the Pleurobrachia consists of a transparent sphere, varying, however, from the perfect sphere in being somewhat oblong, and also by a slight compression on two opposite sides (Figs. 27 and 28), so as to render its horizontal diameter longer in one direction than in the other (Fig. 30). This divergence from the globular form, so slight in Pleurobrachia as to be hardly perceptible to the casual observer, establishing two diameters of different lengths at right angles with each other, is equally true of the other genera. It is interesting and important, as showing the tendency in this highest group of Acalephs to assume a bilateral character. This bilaterality becomes still more marked in the highest class of Radiates, the Echinoderms. Such structural tendencies in the lower animals, hinting at laws to be more fully developed in the higher forms, are always significant, as showing the intimate relation between all parts of the plan of creation. This inequality of the diameters is connected with the disposition of parts in the whole structure, the locomotive fringes and the vertical tubes connected with them being arranged in sets of four on either side of a plane passing through the longer diameter, showing thus a tendency toward the establishment of a right and left side of the body, instead of the perfectly equal disposition of parts around a common centre, as in the lower Radiates.

The Pleurobrachia are so transparent, that, with some preparatory explanation of their structure, the most unscientific observer may trace the relation of parts in them. At one end of the sphere is the transverse split (Fig. 27), that serves them as a mouth; at the opposite pole is a small circumscribed area, in the centre of which is a dark eye-speck. The eight rows of locomotive fringes run from pole to pole, dividing the whole surface of the body like the ribs on a melon. (Figs. 27, 28.) Hanging from either side of the body, a little above the area in which the eye-speck is placed, are two most extraordinary appendages in the shape of long tentacles, possessing such wonderful power of extension and contraction that, while at one moment they may be knotted into a little compact mass no bigger than a pin's head, drawn up close against the side of the body, or hidden within it, the next instant they may be floating behind it in various positions to a distance of half a yard and more, putting out at the same time soft plumy fringes (Fig. 29) along one side, like the beard of a feather. One who has never seen these animals may well be pardoned for doubting even the most literal and matter-of-fact account of these singular tentacles. There is no variety of curve or spiral that does not seem to be represented in their evolutions. Sometimes they unfold gradually, creeping out softly and slowly from a state of contraction, or again the little ball, hardly perceptible against the side of the body, drops suddenly to the bottom of the tank in which the animal floating, and one thinks for a moment, so slight is the thread-like attachment, that it has actually fallen from the body; but watch a little longer, and all the filaments spread out along the side of the thread, it expands to its full length and breadth, and resumes all its graceful evolutions.

One word of the internal structure of these animals, to explain its relation to the external appendages. The mouth opens into a wide digestive cavity (Figs. 27, 28), enclosed between two vertical tubes. Toward the opposite end of the body these tubes terminate or unite in a single funnel-like canal, which is a reservoir as it were for the circulating fluid poured into it through an opening in the bottom of the digestive cavity. The food in the digestive cavity becomes liquefied by mingling with the water entering with it at the mouth, and, thus prepared, it passes into this canal, from which, as we shall presently see, all the circulating tubes ramifying throughout the body are fed. Two of these circulating tubes, or, as they are called from the nature of the liquid they contain, chymiferous tubes, are very large, starting horizontally and at right angles with the digestive cavity from the point of junction between the vertical tubes (Fig. 30) and the canal. Presently they give off two branches, those again ramifying in two directions as they approach the periphery, so that each one of the first main tubes has multiplied to four, before its ramifications reach the surface, thus making in all eight radiating tubes. So far, these eight tubes are horizontal, all diverging on the same level; but as they reach the periphery each one gives rise to a vertical tube, running along the surface of the body from pole to pole, just within the rows of locomotive fringes on the outer surface, and immediately connected with them (Figs. 27, 28). As in all the Ctenophorae, these fringes keep up a constant play of color by their rapid vibrations. In Pleurobrachia the prevailing tint is a yellowish pink, though it varies to green, red, and purple, with the changing motions of the animal. We have seen that the vertical tubes between which the digestive cavity is enclosed, start like the cavity itself from that pole of the body where the mouth is placed, and that, as they approach the opposite pole, at a distance from the mouth of about two thirds the whole length of the body, they unite in the canal, which then extends to the other pole where the eye-speck is placed. As it is just at this point of juncture between the tubes and the canal that the two main horizontal tubes arise from which all the others branch on the same plane (Figs. 27, 28), it follows that they reach the periphery, not on a level with the pole opposite the mouth, but removed from it by about one third the height of the body. In consequence of this the eight vertical tubes arising from the horizontal ones, in order to run the entire length of the body from pole to pole, extend in opposite directions, sending a branch to each pole, though the branch running toward the mouth is of course the longer of the two. The tentacles have their roots in two sacs within the body, placed at right angles with the split of the mouth. (Figs. 27, 30.) They open at the surface on the opposite side from the mouth, though not immediately within the area at which the eye-speck is placed, but somewhat above it, and at a little distance on either side of it. The tentacles may be drawn completely within these sacs, or be extended outside, as we have seen, to a greater or less degree, and in every variety of curve or spiral.

Bolina. (Bolina alata AG.)

The Bolina (Fig 32), like the Pleurobrachia, is slightly oval in form, with a longitudinal split at one end of the body, forming a mouth which opens into a capacious sac or digestive cavity. But it differs from the Pleurobrachia in having the oral end of the body split into two larger lobes (Fig. 31), hanging down from the mouth. These lobes may gape widely, or they may close completely over the mouth so as to hide it from view, and their different aspects under various degrees of expansion or contraction account for the discrepancies in the description of these animals. We have seen that the Pleurobrachia moves with the mouth upward; but the Bolina, on the contrary, usually carries the mouth downward, though it occasionally reverses its position, and in this attitude, with the lobes spread open, it is exceedingly graceful in form, and looks like a white flower with the crown fully expanded. These broad lobes are balanced on the other sides of the body by four smaller appendages, divided in pairs, two on each side (Fig. 32), called auricles. These so-called auricles are in fact organs of the same kind as the larger lobes, though less developed. The rows of locomotive flappers on the Bolina differ in length from each other (Fig. 31), instead of being equal, as in the Pleurobrachia. The four longest ones are opposite each other on those sides of the body where the larger lobes are developed, the four short ones being in pairs on the sides where the auricles are placed. At first sight they all seem to terminate at the margin of the body, but a closer examination shows that the circulating tubes connected with the longer row extend into the lobes, where they wind about in a variety of complicated involutions. (Fig. 32.) The movements of the Bolina are more sluggish than those of the Pleurobrachia, and the long tentacles, so graceful an ornament to the latter, are wanting in the former. With these exceptions the description given above of the Pleurobrachia will serve equally well for the Bolina. The structure is the same in all essential points, though it differs in the size and proportion of certain external features, and its play of color is less brilliant than that of the Pleurobrachia. The Bolina, with its slow, undulating motion, its broad lobes sometimes spreading widely, at other times folded over the mouth, its delicacy of tint and texture, and its rows of vibrating fringes along the surface, is nevertheless a very beautiful object, and well rewards the extreme care without which it dies at once in confinement.

Idyia. (Idyia roseola AG.)

The lowest genus of Ctenophorae found on our coast, the Idyia (Fig. 33), has neither the tentacles of the Pleurobrachia, nor the lobes of the Bolina. It Is a simple ovate sphere, the interior of which is almost entirely occupied by an immense digestive cavity. It would seem that the reception and digestion of food is intended to be the almost exclusive function of this animal, for it has a mouth whose ample dimensions correspond with its capacious stomach. Instead of the longitudinal split serving as a mouth, in the Bolina and Pleurobrachia, one end of the body in the Idyia is completely open (Fig. 33), so that occasionally some unsuspicious victim of smaller diameter than itself may be seen to swim into this wide portal, when suddenly the door closes behind him with a quick contraction, and he finds himself a prisoner. The

Idyia does not always obtain its food after this indolent fashion however, for it often attacks a Bolina or Pleurobrachia as large or even larger than itself, when it extends its mouth to the utmost, slowly overlapping the prey it is trying to swallow by frequent and repeated contractions, and even cutting off by the same process such portions as cannot be forced into the digestive cavity.

The general internal structure of the Idyia corresponds with that of the Bolina and Pleurobrachia; it has the same tubes branching horizontally from the main cavity, then ramifying as they approach the periphery till they are multiplied to eight in all, each of which gives off one of the vertical tubes connected with the eight rows of locomotive flappers. Opposite the mouth is the eye-speck, placed as in the two other genera, at the centre of a small circumscribed area, which in the Idyia is surrounded by delicate fringes, forming a rosette at this end of the body. These animals are exceedingly brilliant in color; bright pink is their prevailing hue, though pink, red, yellow, orange, green, and purple, sometimes chase each other in quick succession along their locomotive fringes. At certain seasons, when most numerous, they even give a rosy tinge to patches on the surface of the sea. Their color is brightest and deepest before the spawning season, but as this advances, and the ovaries and spermaries are emptied, they grow paler, retaining at last only a faint pink tint. They appear early in July, rapidly attain their maximum size, and are most numerous during the first half of August. Toward the end of August they spawn, and the adults are usually destroyed by the early September storms, the young disappearing at the same time, not to be seen again till the next summer. It is an interesting question, not yet solved, to know what becomes of the summer's brood in the following winter. They probably sink into deep waters during this intervening period. The Idyia, like the Pleurobrachia, moves with the mouth upward, but inclined slightly forward also, so as to give an oblique direction to the axis of the body.[4]

[Footnote 4: Until this summer only the three genera of Ctenophorae above mentioned were supposed to exist along our coast, but during the present season I have had the good fortune to find two additional ones. One of them, the Lesueuria, resembles a Bolina with the long lobes so cut off, that they have a very stunted appearance in comparison with those of the Bolina. The other, the Mertensia, is closely allied to Pleurobrachia; it is exceedingly flattened and pear-shaped. This species was discovered long ago by Fabricius, but had escaped thus far the attention of other naturalists. (A. Agassiz.)]

EMBRYOLOGY OF CTENOPHORAE.

All the Ctenophorae are reproduced from eggs, these eggs being so transparent that one may follow with comparative ease the changes undergone by the young while still within the egg envelope. Unfortunately, however, they are so delicate that it is impossible to keep them alive for any length of time, even by supplying them constantly with fresh sea-water, and keeping them continually in motion, both of which are essential conditions to their existence. It is therefore only from eggs accidentally fished up at different stages of growth that we may hope to ascertain any facts respecting the sequence of their development. When hatched, the little Ctenophore is already quite advanced. It is small when compared with the size of the egg envelope, and long before it is set free, it swims about with great velocity within the walls of its diminutive prison (Fig. 35). The importance of studying the young stages of animals can hardly find a better illustration than among the Ctenophorae. Before their extraordinary embryonic changes were understood, many of the younger forms had found their way into our scientific annals as distinct animals, and our nomenclature thus became burdened with long lists of names which will disappear as our knowledge advances.

The great size of their locomotive flappers in proportion to the rest of the body, is characteristic of the young Ctenophorae. They seem like large paddles on the sides of these tiny transparent spheres, and, owing to their great power as compared with those of the adult, the young move with extraordinary rapidity. The Pleurobrachia alone retains its quickness of motion in after life, and although its long graceful streamers appear only as short stumpy tentacles in the young (Fig. 34), yet its active little body would be more easily recognized in the earlier stages of growth than that of the other Ctenophorae. Figs. 34, 35, and 36 show the Pleurobrachia at various stages of growth; Fig. 34, with its thick stunted tentacles and short rows of flappers, is the youngest; the flappers themselves are rather long at this age, looking more like stiff hairs than like the minute fringes of the adult. In Fig. 35 the tentacles are already considerably longer and more delicate; in Fig. 36 the vertical tubes are already completed, while Figs. 27-29 present it in its adult condition.

The Idyia differs greatly in appearance at different periods of its development, and, indeed, no one would suspect, without some previous knowledge of its transformations, that the young Idyia, with its rapid gyrations, its short ambulacral tubes, like immense pouches (Fig. 37), its large pigment spots scattered over the surface (Fig. 38), was an earlier stage of the rosy-hued Idyia, which glides through the water with a scarcely perceptible motion. Figs. 37-40 represent the various stages of its growth. It will be seen how very short are the locomotive fringes (Fig. 39) in comparison with those of the full-grown ones (Fig. 33). It is only in the adult Idyia that these rows attain their full height, and the tubes, ramifying throughout the body (Fig. 40), are completed.

The Bolina, in its early condition, recalls the young Pleurobrachia. At this period it has the same rapid motion, and when somewhat more advanced, long tentacles, resembling those of the Pleurobrachia, make their appearance (Fig. 41); it is only at a later period that the tentacles become contracted, while the large lobes (Fig. 42), so characteristic of Bolina, are formed by the elongation of the oral end of the body, the auricles becoming more conspicuous at the same time (Fig. 43). A little later the lobes enlarge, the movements become more lazy; it assumes both in form and habits the character of the adult Bolina.

The series of changes through which the Ctenophorae pass are as remarkable as any we shall have occasion to describe, though not accompanied with such absolutely different conditions of existence. The comparison of the earlier stages of life in these animals with their adult condition is important, not only with reference to their mode of development, but also because it gives us some insight into the relative standing of the different groups, since it shows us that certain features, permanent in the lower groups, are transient in the higher ones. A striking instance of this occurs in the fact mentioned above, that though the long tentacles so characteristic of the adult Pleurobrachia exist in the young Bolina, they yield in importance at a later period to the lobes which eventually become the predominant and characteristic feature of the latter.

* * * * *

DISCOPHORAE.

The disk of the Discophorae is by no means so delicate as that of the other Jelly-fishes. It seems indeed quite solid, and somewhat like cartilage to the touch, and yet so large a part of its bulk consists of water, that a Cyanea, weighing when alive about thirty-four pounds, being left to dry in the sun for some days, was found to have lost about 99/100 of its original weight,--only the merest film remaining on the paper upon which it had been laid. The prominence of the disk in this group of Jelly-fishes is well characterized by their German name, "Scheiben quallen," viz. disk-medusae. We shall see hereafter that the disk, so large and seemingly solid in the Discophorae, thins out in many of the other Jelly-fishes, and becomes exceedingly concave. This is especially the case in many of the Hydroid Medusae, where it assumes a bell-shaped form, and is constantly spoken of as the bell. It should be remembered, however, in reading descriptions of these animals, that the so-called bell is only a modified disk, and perfectly homologous with that organ in the Discophorae.

The Discophorous Medusae are distinguished from all others by the peculiar nature of the reproductive organs. They are contained in pouches (Fig. 50, o, o, o, o), the contents of which are first discharged into the main cavity, and then pass out through the mouth. Pillars support the four angles of the digestive cavity, thus separating the lower from the upper floor of the disk, while the chymiferous tubes (Fig. 50) branch and run into each other near the periphery, forming a more or less complicated anastomosing network, instead of a simple circular tube, as is the case with the Hydroid Medusae. (Fig. 74.)

Cyanea. (Cyanea arctica PER. et LES.)

In our descriptions of the Discophorae, we may give the precedence to the Cyanea on account of its size. This giant among Jelly-fishes is represented in Fig. 44. It is much to be regretted that these animals, when they are not so small as to escape attention altogether, are usually seen out of their native element, thrown dead or dying on the shore, a mass of decaying gelatinous matter. All persons who have lived near the sea are familiar with the so-called Sea-blubbers, sometimes strewing the sandy beaches after the autumn storms in such numbers that it is difficult to avoid them in walking or driving.

In such a condition the Cyanea is far from being an attractive object; to form an idea of his true appearance, one must meet him as he swims along at midday, rather lazily withal, his huge semi-transparent disk, with its flexible lobed margin, glittering in the sun, and his tentacles floating to a distance of many yards behind him. Encountering one of those huge Jelly-fishes, when out in a row-boat one day, we attempted to make a rough measurement of his dimensions upon the spot. He was lying quietly near the surface, and did not seem in the least disturbed by the proceeding, but allowed the oar, eight feet in length, to be laid across the disk, which proved to be about seven feet in diameter. Backing the boat slowly along the line of the tentacles, which were floating at their utmost extension behind him, we then measured these in the same manner, and found them to be rather more than fourteen times the length of the oar, thus covering a space of some hundred and twelve feet. This sounds so marvellous that it may be taken as an exaggeration; but though such an estimate could not of course be absolutely accurate, yet the facts are rather understated than overstated in the dimensions here given. And, indeed, the observation was more careful and precise than the circumstances would lead one to suppose, for the creature lay as quietly, while his measure was taken, as if he had intended to give every facility for the operation. This specimen was, however, of unusual size; they more commonly measure from three to five feet across the disk, while the tentacles may be thirty or forty feet long. The tentacles are exceedingly numerous (see Fig. 44), arising in eight distinct bunches, from the margin of the disk, and hanging down in a complete labyrinth.

These animals are not so harmless as it would seem, from their soft, gelatinous consistency; it is no pleasant thing when swimming or bathing to become entangled in this forest of fine feelers, for they have a stinging property like nettles, and may render a person almost insensible, partly from pain, and partly from a numbness produced by their contact, before he is able to free himself from the network in which he is caught. The weapons by which they produce these results seem so insignificant, that one cannot but wonder at their power. The tentacles are covered by minute cells, lasso-cells as they are called, (similar to those of Astrangia, Fig. 19,) each one of which contains a whip finer than the finest thread, coiled in a spiral within it.

These myriad whips can be thrown out at the will of the animal, and really form an efficient galvanic battery. Behind the veil of tentacles, and partly hidden by it, four curtains, with lobed or ruffled margins, dimly seen in Fig. 44, hang down from the under surface of the disk. The ovaries are formed by four pendent pouches, placed near the sides of the mouth, and attached to four cavities within the disk. Around the circumference of the disk are eight eye-specks, each formed by a small tube protected under a little lappet or hood rising from the upper surface of the disk. The prevailing color of this huge Jelly-fish is a dark brownish-red, with a light, milk-white margin, tinged with blue, the tentacles and other pendent appendages having a somewhat different hue from the disk. The ovaries are flesh-colored, the curtain formed by the expansion of the lobes of the mouth is dark brown, while the tentacles are of different colors, some being yellow, others purple, and others reddish brown or pink.

Strange to say, this gigantic Discophore is produced by a Hydroid measuring not more than half an inch in height when full grown; could we follow the history of any egg laid by one of these Discophorae in the autumn, and this has indeed been partially done, we should see that, like any other planula, the young hatched from the egg is at first spherical, but presently becomes pear-shaped, and attaches itself to the ground. From the upper end tentacles project (see Fig. 45), growing more numerous, as in Fig. 46, though they never exceed sixteen in number. As it increases in height constrictions take place at different distances along its length, every such constriction being lobed around its margin, till at last it looks like a pile of scalloped saucers or disks strung together (see Fig. 47). The topmost of these disks falls off and dies; but all the others separate by the deepening of the constrictions, and swim off as little free disks (Fig. 48), which eventually grow into the enormous Jelly-fish described above. These three phases of growth, before the relation between them was understood, have been mistaken for distinct animals, and described as such under the names of Scyphistoma, Strobila, and Ephyra.

Aurelia. (Aurelia flavidula PER. et LES.)

Another large Discophore, though by no means to be compared to the Cyanea in size, is our common Aurelia (Figs. 49, 50) Its bluish-white disk measures from twelve to fifteen inches in diameter, but its dimensions are not increased by the tentacles, which have no great power of contraction and expansion, and form a short fringe around its margin, widening and narrowing slightly as the tentacles are stretched or drawn in. It is quite transparent, as may be seen in Fig 49, where all the fine ramifications of the chymiferous tubes as well as the ovaries, are seen through the vault of the disk. Fig. 50 represents the upper surface, with the ovaries around the mouth, occupying the same position as those of the Cyanea, though they differ from the latter in their greater rigidity, and do not hang down in the form of pouches. The males and females in this kind of Jelly-fish may be distinguished by the difference of color in the reproductive organs, which are rose-colored in the males, and of a dull yellow in the females. The process of development is exactly the same in the Aurelia as in the Cyanea, though there is a very slight difference in their respective Hydroids. They are, however, so much alike, that one is here made to serve for both, the above figures being taken from the Hydroid of the Aurelia. It is curious, that while, as in the case of the Aurelia and Cyanea, very dissimilar Jelly-fishes may arise from almost identical Hydroids, we have the reverse of the proposition, in the fact that Hydroids of an entirely distinct character may produce similar Jelly-fishes.

The embryos or little planulae, hatched from the Cyanea and Aurelia in the fall, seem to be gregarious in their mode of life, swimming about together in great numbers till they find a suitable point of attachment, and assume their fixed Hydroid existence. The Cyaneae, however, when adult, are usually found singly, while the Aureliae, on the contrary, seek each other, and commonly herd together.

The Campanella. (Campanella pachyderma A. AG.)

The Campanella (Fig. 51) is a pretty little Jelly-fish, not larger than a pin's head, reproduced directly from eggs, without passing through the Hydroid stage. During its early stages of growth it probably remains attached to floating animals, thus leading a kind of parasitic existence; but as its habits are not accurately known, this cannot be asserted as a constant fact respecting them. The veil in this Jelly-fish is very large, forming pendent pouches hanging from the circular canal (see Fig. 51), and leaving but just room enough for the passage of the proboscis between the folds. It may not be amiss to introduce here a general account of this organ, which occurs in many of the Medusae, though it has very different proportions in the various kinds. It is a delicate membrane, hanging from the circular tube, so as partially to close the mouth of the bell, leaving a larger or smaller opening for the passage of the water, which is taken in and forced out again by the alternate expansions and contractions of the bell.

There are but four chymiferous tubes in the Campanella, and four stiff tentacles, which in consequence of the peculiar character of the veil appear, when the animal is seen in profile, to start from the middle of the disk. The ovaries consist of eight pouches, placed near the point of junction of the four chymiferous tubes. (Fig. 52.) This little Medusa is of a dark yellowish color with brownish ocellated spots, scattered profusely over the upper part of the disk.

Circe. (Trachynema digitale A. AG.)

Among the Jelly-fishes, the position of which is somewhat doubtful, is the Circe (Fig. 53), differing greatly in outline from the ordinary Jelly-fishes. As may be seen in Figure 53, the bell forms but a small portion of the animal; it rises in a sharp cone on the summit, thinning out at the lower edge, to form the large cavity in which hangs the long proboscis and the eight ovaries, four of which may be seen in Fig. 53 crowded with eggs. The Circe differs in consistency, as well as in form, from other Jelly-fishes. It is hard and horny to the touch, and the veil, usually so light and filmy, is here a thick folded membrane, which at every stroke of the animal forces the water in and out of the cavity. It is very active, moving by powerful jerks, each one of which throws it far on its way. It advances usually in straight lines; or, if it wishes to change its direction, it drives the water out of the veil suddenly on one side or the other, so as to shoot off, sometimes at right angles with its former path. Four large pedunculated eyes, hidden in the figure by the tentacles, stand out prominently from the circular tube. When the animal is in motion, the tentacles are carried closely curled around the edge of the disk, as in Fig. 53, where the Circe is represented under a magnifying power of two and a half diameters. This Jelly-fish is of a delicate rose color, the tentacles assuming, however, a dark-purple tint at their extremities when contracted.

Lucernaria. (Haliclystus auricula CLARK.)

One of the prettiest and most graceful, as well as one of the most common of our Jelly-fishes, is the Lucernaria (Fig. 54). It has such an extraordinary contractility of all its parts, that it is not easy to describe it under any definite form or position, since both are constantly changing; but perhaps of all its various attitudes and outlines none are more normal to it than those given in Fig. 54. It frequently raises itself in the upright position represented here by the individual highest on the stem, spreading itself in the form of a perfectly symmetrical cup or vase, the margin of which is indented by a succession of inverted scallops, the point of junction between every two scallops being crowned by a tuft of tentacles. But watch it for a while, and the sides of this vase turn backward, spreading completely open, till they present the whole inner surface, with the edges even curved a little downward, drooping slightly, and the proboscis rising in the centre. In such an attitude one may trace with ease the shape of the mouth, the lobes surrounding it, as well as the tubes and cavities radiating from it toward the margin. A touch is, however, sufficient to make it close upon itself, shrinking together in the attitude of the third individual in Fig. 54, or even drawing its tentacles completely in, and contracting all its parts till it looks like a little ball hanging on the stem. These are but a few of its manifold changes, for it may be seen in every phase of expansion and contraction. Let us now look for a moment at the details of its structure. The resemblance to a cup or vase, as in the upper figure of the wood-cut (Fig. 54), is deceptive; for a vase is hollow, whereas the Lucernaria, though so delicate and transparent that its upper surface, when thus stretched, seems like a mere film, is nevertheless a solid gelatinous mass, traversed by certain channels, cavities, and partitions, but otherwise continuous throughout. The peduncle by which it is attached is but an extension of the floor of a gelatinous disk, corresponding to that of any Jelly-fish. Four tubes pass through the whole length of this peduncle, and open into four chambers, dividing the digestive cavity above into as many equal spaces. (Fig. 55.) These spaces are produced by folds in the upper floor of the disk, uniting it to the lower floor at given distances, and forming so many partition-walls, dividing the digestive sac into four distinct cavities. These lines of juncture between the two floors, where the partitions occur, produce the four radiating lines, running from the proboscis to the margin of the disk, on the upper surface. (Fig. 55.) The triangular figures, running from the mouth to each cluster of tentacles, are produced by the ovaries, which consist of distinct pouches or bags attached to the upper surface of the disk, and hanging down into the cavities below; every little dot within these triangular spaces represents such a bag. Each bag is crowded with eggs, which drop into the digestive cavity at the spawning season, and are passed out at the mouth. The tentacles always grow in clusters, but are nevertheless arranged according to a regular order. They are club-shaped at their extremities, but are hollow throughout, opening into the chambers of the digestive cavity, two of the clusters thus being connected with each chamber. Their chief office seems to be to catch the food and convey it to the mouth, though they may also be used as a kind of suckers, and the animal not unfrequently attaches itself by means of these appendages. Between every two clusters of tentacles will be observed a short, single appendage, of an entirely different appearance. These are the so-called auricles, and though so unlike tentacles in the adult animal, when in their earlier stages (Fig. 56) they resemble each other closely. But as their development goes on, the tentacles stretch out into longer, more delicate flexible organs, while the auricles remain short and compact throughout life. They contain a slight pigment spot representing an eye, though how far it serves the purpose of vision remains doubtful. They are chiefly used by the animal as a means of adhering to any surface upon which it may wish to fasten itself; for the Lucernaria, though usually found attached to eel-grass in shoal water, has the power of independent motion, and frequently separates from its resting-place, floating about freely in the water for a while, or attaching itself anew by means of the auricles and tentacles upon some other object. The color of this pretty Acaleph varies from a greenish hue to green, with a faint tinge of red, or to a reddish brown. One of its commonest and most exquisite tints is that of a pale aqua-marine. It may be found along our shores wherever the eel-grass grows, and as far out as this plant extends. It thrives admirably in confinement, and for this reason is especially adapted to the aquarium.

* * * * *

HYDROIDS.

Under this order, the general character of which has already been explained in the introductory chapter on Acalephs, are included a number of groups which, whether as Hydroid communities in their earlier phases of existence, or as free swimming Medusae in their farther development, challenge our admiration, both for their beauty of form and color, and their grace of motion. Some of them are so minute that they escape the observation of all but those who are laboriously seeking for the hidden treasures of the microscopic world, but the greater number are large enough to be readily found by the most inexperienced collector, when his attention is once drawn to them; and he may easily stock his aquarium with these pretty little communities, and even trace the development of the Jelly-fishes upon them.

To the Hydroids belong the Campanularians, the Sertularians, and the Tubularians. Some examples of each, as represented on our shores, will be found under their different heads, accompanied with full descriptions. There is another group usually considered as distinct from Hydroids, and known as a separate order among Acalephs, under the name of Siphonophorae, but included with them here in accordance with the views of Vogt, Agassiz, and others, in whose opinion they differ from the ordinary Hydroid communities only in being free and floating, instead of fixed to the ground. Some new facts, published here for the first time, tend to sustain the accuracy of this classification.[5] With these few preliminary remarks to show the connection of the order, let us now look at some of the animals belonging to it more in detail.

[Footnote 5: See Chapter on Nanomia.]

Campanularians.

All the Campanularians, of which Oceania (Fig. 68), Clytia (Fig. 73), and Eucope (Fig. 61) form a part, belong among those little shrub-like communities of animals called Hydroids, from which most of our Jelly-fishes are developed. They differ in one essential feature from the Tubularians. (Fig. 93.) The whole stem, from summit to base, is enveloped in a horny sheath, extending around both the fertile and sterile individuals of the community, and forming a network at the base of the stem, which serves as a kind of foundation for the whole stock. To the naked eye such a community looks like a tiny shrub (see Fig. 57), with the branches growing in regular alternation on either side of the stems. The reproductive calycles, i.e. the protecting envelopes covering the young Medusae, usually arise in the angles of the branches formed by a prolongation of the sheath. These calycles or bells, as they are called, assume a great variety of shapes,--elliptical, round, pear-shaped, or ringed like the Clytia. (Fig. 72.) In one such bell there may be no less than twenty or thirty Medusae developed one below the other; when ready to hatch, the calycle bursts and allows them to escape.

Eucope. (Eucope diaphana AG.)

In Figs. 60 and 61 we have a representation of our little Eucope, one of the prettiest of the Jelly-fishes belonging to this group; Fig. 57 represents the Hydroid from which it arises; a single branch with the reproductive bell being magnified in Fig. 58. In Fig. 59 is seen a portion of the Jelly-fish disk, with the fringe of tentacles highly magnified. The disk of the Eucope (Fig. 60) looks like a shallow bell, of which the proboscis often seems to form the handle; for the disk has such an extraordinary thinness that it turns inside out with the greatest ease, so that the inner surface may become at any moment the outer one, with the proboscis projecting from it, as in Fig. 60, while the next movement of the animal may reverse its whole position, and the proboscis then hangs down from the inside, as in other Jelly-fishes. (See Fig. 61.)

The tentacles are solid and stiff like little hairs, and two of them, in each quarter-segment of the disk, have small concretions at the base, which are no doubt eye-specks. (See Fig. 62.) Along the chymiferous tubes little swellings are developed, which increase gradually, and become either ovaries or spermaries, according to the sex of the animal. (Fig. 63.) In the adult the genital organs hang down, like elongated bags, from the chymiferous tubes. (Fig. 64.) The tentacles are numerous, multiplying to about a hundred and ninety-two in the adult, and increasing according to the numerical law to be explained in the description of the Oceania.

This little Jelly-fish is one of the most common in our Bay. There is not a night or day when they cannot be taken in large numbers, from the early spring till late in the autumn; and as the breeding season lasts during the whole of that period, they are found in all possible stages of growth. In consequence of this, the course of their development, and the relation between the different phases of their existence as Hydroids, and afterwards as Acalephs, are well known, though the successive steps of their growth have not been traced connectedly, as in some of the other Jelly-fishes, the Tima or Melicertum, for instance. The process is, however, so similar throughout the class of Hydroids, that, having followed it from beginning to end in some of the groups, we have the key to the history of others, whose development has not been so fully traced. The eggs laid by the Eucope in the autumn develop into planulae, which acquire their full size as Hydroid communities toward the close of the winter, and the development of the young Medusae upon them, as described above, begins with the opening spring.

Oceania. (Oceania languida A. AG.)

The Oceania (Fig. 68) is so delicate and unsubstantial, that with the naked eye one perceives it only by the more prominent outlines of its structure. We may see the outline of the disk, but not the disk itself; we may trace the four faint thread-like lines produced by the radiating tubes traversing the disk from the summit to the margin; and we may perceive, with far more distinctness, the four ovaries attached to these tubes near their base; we may see also the circular tube uniting the radiating tubes, and the tentacles hanging from it, and we can detect the edge of the filmy veil that fringes the margin of the disk. But the substance connecting all these organs is not to be distinguished from the element in which it floats, and the whole structure looks like a slight web of threads in the water, without our being able to discern by what means they are held together. Under the microscope, however, the invisible presently becomes visible, and we find that this Jelly-fish, like all others, has a solid gelatinous disk.

Let us begin with its earlier condition. When it first escapes from the parent Hydroid stock, the Oceania is almost spherical in form. (See Fig. 65.) The disk is divided by four chymiferous tubes, running from the summit to the margin, where they meet the circular tube in which they all unite. At this time, it has but two well-developed tentacles, opposite each other on the margin of the disk, just at the base of two of the chymiferous tubes (Fig. 66), while two others are just discernible in a rudimentary state, forming slight projections at the base of the two other tubes. Fig. 66 gives a view of the animal from below, at this stage of its growth, while Fig. 65 shows it in profile. It will be seen by the latter how very spherical is the outline of the disk at this period, while the proboscis, in which are placed the mouth and digestive cavity, is quite long, and hangs down considerably below the lower surface of the disk. As the animal advances in age the disk loses its spherical outline, and becomes much flattened, as may be seen in Fig. 67. It may be well to introduce here some explanation of the law according to which the different sets of tentacles follow each other in successive cycles of growth, since it is a law of almost universal application in Jelly-fishes and Polyps; and, owing to the smaller number and simpler arrangement of the tentacles in Oceania, it may be more easily analyzed in them than in many others, where the number and complication of the different sets of tentacles make it very difficult to trace their relation to each other during their successive growth. We have seen that the Oceania begins life with only two tentacles. These form the first set, and are marked with the number 1 in the subjoined diagram, which gives the plan of all the different sets in their regular order. The second set, marked 2, consists also of two, which are developed at equal distances between the first two, i.e. at right angles with them. The third set, however, marked 3, consists of four, as do all the succeeding sets, and they are developed between the first and second. The fourth set comes in between the first and third; the fifth between the third and second; the sixth between the first and fourth; the seventh between the fifth and second; the eighth between the third and fourth; the ninth between the fifth and third. The ultimate number of tentacles in the Oceania is thirty-two, or sometimes thirty-six, and the cycles always in twos or multiples of two. But whatever be the number included in the successive sets of tentacles, and the unit for the first set ranges from two to forty-eight, the law in different kinds of Jelly-fishes is always the same, the youngest set always forming between the oldest preceding set. Thus the fourth set comes in between the first and third, and the fifth between the second and third, the intervals occupied now by the fourth set, being limited by the first set of tentacles on one side, and by the third set on the other side, while the intervals occupied by the fifth set are bounded by the second and third sets.

The little spheres represented between the tentacles on the margin of the disk, in Figs. 65-67, are eye-specks, and these continue to increase in number with age; in this the Oceania differs from the Eucope, in which it will be remembered there were but two eye-specks in each quarter-segment of the disk throughout life. Fig. 68 represents the adult Oceania in full size, when it averages from an inch and a half to two inches in diameter. It is slow and languid in its movements, coming to the surface only in the hottest hours of the summer days; at such times it basks in the sun, turning lazily about, and dragging its tentacles after it with seeming effort. Sometimes it remains for hours suspended in the water, not moving even its tentacles, and offering a striking contrast to its former great activity when young, and to the lively little Eucope, which darts through the water at full speed, hardly stopping to rest for a moment. If the Oceania be disturbed it flattens its disk, and folds itself up somewhat in the shape of a bale (see Fig. 69), remaining perfectly still, with the tentacles stretching in every direction. When the cause of alarm is removed, it gently expands again, resuming its natural outline and indolent attitudes. The number of these animals is amazing. At certain seasons, when the weather is favorable, the surface of the sea may be covered with them, for several miles, so thickly that their disks touch each other. Thus they remain packed together in a dense mass, allowing themselves to be gently drifted along by the tide till the sun loses its intensity, when they retire to deeper waters. Some points, not yet observed, are still wanting to complete the history of this Jelly-fish. By comparing such facts, however, as are already collected respecting it, with our fuller knowledge of the same process of growth in the Eucope, Tima, and Melicertum, we may form a tolerably correct idea of its development. It is hatched from a Campanularia.

Clytia. (Clytia bicophora AG.)

In Figs. 70-73 we have the Acalephian and Hydroid stages of the Clytia (Fig. 73), another very pretty little Jelly-fish, closely allied to the Oceania. When first hatched, like the Oceania, it is very convex, almost thimble-shaped (see Fig. 70), but a little later the disk flattens and becomes more open, as in Fig. 71. In Fig. 72, we have a branch of the Hydroid, a Campanularia, greatly magnified, with the annulated reproductive calycle attached to it, and crowded with Jelly-fishes ready to make their escape as soon as the calycle bursts. The adult Clytia (Fig. 73) is somewhat smaller and more active than the Oceania, and is easily recognized by the black base of its tentacles, at their point of juncture with the margin of the disk. It is more commonly found at night, than in the day-time, being nocturnal in its habits.

Zygodactyla. (Zygodactyla groenlandica AG.)

Little has been known, and still less published, of this remarkable genus of Jelly-fish (Figs. 74, 75) up to the present time. The name Zygodactyla, or Twinfinger, was given to it by Brandt, from drawings made by Mertens, who had some opportunity of studying it in his journey around the world. These drawings were published in the Transactions of the St. Petersburg Academy. In the year 1848 Professor Agassiz read a paper upon one of the species of this genus belonging to our coast, before the American Academy, in which he called it Rhacostoma, not being aware that it had already received a name, and gave some account of its extraordinary phosphorescent properties. The name Rhacostoma must of course yield to that of Zygodactyla, which has a prior claim.

The average size of this Jelly-fish when full grown is from seven to eight inches in diameter; sometimes it may measure even ten or eleven, but this is rather rare. The light-violet colored disk is exceedingly delicate and transparent, its edge being fringed with long fibrous tentacles, tinged with darker violet at their point of juncture with the disk, and hanging down a yard and more when fully extended, though they vary in length according to the size of the specimen, and, in consequence of their contractile power, may seem much shorter at some moments than at others. The radiating tubes in this Jelly-fish are exceedingly numerous, the whole inner surface of the disk being ribbed with them. (See Figs. 74 and 75.) The ovaries follow the length of the tubes, though they do not extend quite to their extremity, where they join the circular tube around the margin of the disk; nor do they start exactly at the point where the tubes diverge from the central cavity, but a little below it. (Fig. 74.) Each ovary consists of a long, brownish, flat bag, split along the middle, so closely folded together that it seems like a flat blade attached along the length of the tube. Perhaps a better comparison would be to a pea-pod greatly elongated, with the edges split along their line of juncture, and attached to a tube of the same length. The ovaries are not perfectly straight, but slightly waving, as may be seen in Fig. 74, and these undulations are stronger when the ovaries are crowded with eggs, as is the case at the time of spawning.

The large digestive cavity hangs from the centre of the under side of the disk (Fig. 75), terminating in the proboscis, which, in this kind of Jelly-fish, is short in proportion to the diameter of the disk, while the opening of the mouth is very large. (Fig. 74.) It is unfortunate that a variety of inappropriate names, likely to mislead rather than aid the unscientific observer, have been applied to different parts of the Jelly-fish. What we call here digestive cavity, proboscis, and mouth, are, in fact, parts of one organ. An exceedingly delicate, transparent, filmy membrane hangs from the under side of the disk; that membrane forms the outer wall of the digestive cavity, which it encloses; it narrows toward its lower margin, leaving open the circular aperture called the mouth; this narrowing of the membrane is produced by a number of folds in its lower part, while at its margin these folds spread out to form ruffles around the edge of the mouth, and these ruffles again extend into the long scalloped fringes hanging down below.

The motion of these Jelly-fishes is very slow and sluggish. Like all their kind, they move by the alternate dilatation and contraction of the disk, but in the Zygodactyla these undulations have a certain graceful indolence, very unlike the more rapid movements of many of the Medusae. It often remains quite motionless for a long time, and then, if you try to excite it by disturbing the water in the tank, or by touching it, it heaves a slow, lazy sigh, with the whole body rising slightly as it does so, and then relapses into its former inactivity. Indeed, one cannot help being reminded, when watching the variety in the motions of the different kinds of Jelly-fishes, of the difference of temperament in human beings. There are the alert and active ones, ever on the watch, ready to seize the opportunity as it comes, but missing it sometimes from too great impatience; and the slow, steady people, with very regular movements, not so quick perhaps, but as successful in the long run; and the dreamy, indolent characters, of which the Zygodactyla is one, always floating languidly about, and rarely surprised into any sudden or abrupt expression. One would say, too, that they have their aristocratic circles; for there is a delicate, high-bred grace about some of them quite wanting in the coarser kinds. The lithe, flexible form of the greyhound is not in stronger contrast to the heavy, square build of the bull dog, than are some of the lighter, more frail species of Jelly-fish to the more solid and clumsy ones. Among these finer kinds we would place the Tima. (Fig. 76.)

Tima. (Tima formosa AG.)

One's vocabulary is soon exhausted in describing the different degrees of consistency in the substance of Jelly-fishes. Delicate and transparent as is the Tima, it has yet a certain robustness and solidity beside the Oceania, described above. In fact, all are gelatinous, all are more or less transparent, and it is not easy to describe the various shades of solidity in jelly. Perhaps they may be more accurately represented by the impression made upon the touch than upon the sight. If, for instance, you place your hand upon a Zygodactyla, you feel that you have come in contact with a substance that has a positive consistency; but if you dip your finger into a bowl where a Tima is swimming, and touch its disk, you will feel no difference between it and the water in which it floats, and will not be aware that you have reached it till the animal shrinks away from the contact.

The adult Tima, represented in Fig. 76, is not more than an inch and a half or two inches in diameter. Instead of countless tubes diverging from the digestive cavity to the margin of the disk, as in the Zygodactyla, there are but four. The digestive cavity in the Tima is much smaller than in the Zygodactyla, and is placed at the end of the proboscis, which is long, and hangs down far below the disk. This removal of the digestive cavity to the extremity of the proboscis gives to the tubes arising from it a very different and much sharper curve than they have in the Zygodactyla. In the Tima they start from the end of the proboscis, as may be seen in the wood-cut (Fig. 76), and then turn abruptly off, when they arrive at the under surface of the disk, to reach its margin. The disk has, as usual, its veil and its fringe of tentacles; the tentacles in the full-grown Tima are few,--seven in all the four intermediate spaces between the tubes, with one at the base of each tube, making thirty-two in all. The ovaries, which are milk-white, follow the line of the tubes, as in the Zygodactyla, and have very undulating folds when full of eggs. The tubes meet in the digestive cavity, the margin of which spreads out to form four ruffled edges that hang down from it. One of these ruffles, considerably magnified, is represented in Fig. 77. In Fig. 78 we have a portion of the Hydroid stock from which this Jelly-fish arises, also greatly magnified. The Tima is very active, yet not abrupt in its motions; but when in good condition it is constantly moving about, rising to the surface by the regular pulsations of the disk, or swimming from side to side, or poising itself quietly in the water, giving now and then a gentle undulation to keep itself in position.

Though not a very frequent visitor of our shores, the appearance of the Tima is not limited by the seasons, since they are found at all times of the year. It is a fact, unexplained as yet, that the Tima and many other Jelly-fishes are never seen except when full grown. What may be the haunts and habits of these animals from the time of their hatching till they make their appearance again in the adult condition, is not known, though it is probable that they remain at the bottom during this period, and only come to the surface to spawn. This impression is confirmed by the observations made upon a very young Cyanea which was kept for a long time in confinement; but a question of this kind cannot of course be settled by a single experiment.[6]

[Footnote 6: Since the above was written, I have had an opportunity of learning some additional facts respecting the habits of the young Cyanea, which may, perhaps, apply to other Jelly-fishes also. Having occasion to visit the wharves at Provincetown at about four o'clock one morning, I was surprised to find thousands of the spring brood of Cyaneae, hitherto supposed to pass the early period of their existence wholly in deep water, floating about near the surface. They varied in size, some being no larger than a three-cent-piece, while others were from an inch in diameter to three inches. It would seem that they make their appearance only during the earliest morning hours, for at seven o'clock, when I returned to the same spot, they had all vanished. It may be that other young Medusae have the same habits of early rising, and that instead of coming to bask in the midday sunshine, like their elders, they prefer the cooler hours of the dawn. (A. Agassiz.)]

Melicertum. (Melicertum campanula PER. et LES.)

A pretty Medusa, smaller and far more readily obtained than the Tima, is the Melicertum. (Fig. 80.) Its disk has a yellowish hue, and from its margin hangs a heavy row of yellow tentacles, while the eight ovaries (Fig. 79) are of a darker shade of the same color. This little golden-tinted Jelly-fish, moving through the water with short, quick throbs, produced by the rapid rise and fall of the disk, is a very graceful object. Its bright color, made particularly prominent by the darker undulating lines of the ovaries, which become very marked near the spawning season, renders it more conspicuous in the water than one would suppose from its size; for it does not measure more than an inch in height when full grown. (See Fig. 80.)

Development of Melicertum and Tima.

In the Melicertum and Tima we have had the good fortune to trace the process by which the eggs are changed into Hydroid communities. If any one has a curiosity to follow for themselves this singular history of alternate generations, the Melicertum is a good subject for the experiment, as it thrives well in confinement. After keeping a number of them in a large glass jar for a couple of days at the time of spawning, it will be found that the ovaries, which were at first quite full of eggs, are emptied, and that a number of planulae; are swimming about near the bottom of the vessel. After a day or two the outline of these planulae, spherical at first, becomes pear-shaped (see Fig. 81), and presently they attach themselves by the blunt end to the bottom of the jar. (Fig. 82.) Thus their Hydroid life begins; they elongate gradually, the horny sheath is formed around them, tentacles arise on the upper end, short and stunted at first, but tapering rapidly out into fine flexible feelers, the stem branches, and we have a little Hydroid community (Fig. 83), upon which, in the course of the following spring, the reproductive calycles containing the Medusae buds will be developed, as in the case of the Eucope and Clytia. The Tima passes through exactly the same process, though the shape of the planulae and the appearance of the young differ from that of the Melicertum, as may be seen in Fig. 78, where a single head of the Tima Hydroid, greatly magnified, is represented. By combining the above observations upon the development of the Hydroids of the Melicertum and Tima with those previously mentioned upon the young Medusa arising from reproductive calycles in the Eucope and Clytia, we get a complete picture of all the changes through which any one of these Hydroid Medusae passes, from its Hydroid condition to the moment when it enters upon an independent existence as a free Jelly-fish.

(Laomedea amphora AG.)

The Medusae of the Campanularians are not all free. On the contrary, in many of the species they always remain attached to the Hydroid, never attaining so high a development as the free Medusae, and withering on the stem after having laid their eggs. Such is the Laomedea amphora, quite common on all the bridges connecting Boston with the country, where, on account of the large amount of food brought down from the sewers by the river, they thrive wonderfully, growing to a great size, sometimes measuring from a foot to eighteen inches in height.

Sertularians.

The Sertularians form another group of Hydroids closely allied to the Campanularians, though differing from them in the arrangement of the sterile Hydrae upon the stem. Among these one of the most numerous is the Dynamena (Dynamena pumila Lamx., Fig. 84), which hangs its yellowish fringes from almost every sea-weed above low-water-mark. It is especially thick and luxuriant on the fronds of our common Fucus vesiculosus. The color is usually of a pale yellow, though sometimes it is nearly white, and when first taken from the water it has a glittering look, such as a white frost leaves on a spray of grass. Fig. 84 represents such a cluster in natural size, while Fig. 85 shows a piece of the stem highly magnified, with a reproductive calycle attached to the side of a sterile Hydra stem. Many of these Sertularian Hydroids assume the most graceful forms, hanging like long pendent streamers from the Laminaria, or in other instances resembling miniature trees. One of these tree-like Sertularians (Dyphasia rosacea Ag.), abundant on all rocks in sheltered places immediately below low-water-mark, is represented in Fig. 86. In both these Sertularians the Medusae wither on the stock, never becoming free. The free Medusae of the Sertularians are only known in their adult condition in a single genus, which is closely allied to Melicertum, and which is produced from a Hydroid genus called Lafoea. Fig. 87 represents one of these young Sertularian Medusae (Lafoea cornuta Lamx.).

Tubularians.

In the Sertularian and Campanularian Hydroids we have found that the communities consist generally of a large number of small individuals, so small, indeed, that it is hardly possible at first glance to distinguish the separate members of these miniature societies. Among the Tubularians, on the contrary, the communities are usually composed of a small number of comparatively large individuals; and indeed these Hydroids may even grow singly, as in the case of the Hybocodon (Fig. 104), which attains several inches in height. There is also another general feature in which the Tubularians differ from both the other groups of Hydroids. In the latter, the horny sheath which encloses the stem extends to form a protecting calycle around the Hydra heads. This protecting calycle is wanting round the heads of the Tubularians, though their stems are surrounded by a sheath.

Sarsia. (Coryne mirabilis AG.)

Among the most common of our Tubularians is a small, mossy Hydroid (Fig. 88), covering the rocks between tides, in patches of several feet in diameter. Fig. 89 represents a single head from this little mossy tuft greatly magnified, in which is seen the medusa bud arising from the stem by the process already described in the other Hydroids. In Fig. 90 we have the little Jelly-fish in its adult condition, about the size of a small walnut, with a wide circular opening, through which passes the long proboscis, hanging from the under surface of the disk to a considerable distance below its margin. The four tentacles are of an immense length when compared to the size of the animal. As a general thing, the tentacles are less numerous in the Tubularian Medusae than in those arising from other Hydroids; they want also the singular limestone concretions found at the base of the tentacles in the Campanularian Medusae. In Fig. 91 we have one of the Tubularian Medusae (Turris vesicaria A. Ag.) which lifts a rather larger number of tentacles than is usual among these Jelly-fishes. We never find the tentacles multiplying almost indefinitely in them, as in Zygodactyla and Eucope. The little Jelly-fish described above is known as Sarsia, while its Hydroid is called Coryne. These names having been given to the separate phases of its existence before their connection was understood, and when they were supposed to represent two distinct animals. They are especially interesting with reference to the history of Hydroids in general, because they were among the first of these animals in whom the true relation between the different phases of their existence was discovered. Lesson named the Sarsia after the great Norwegian naturalist, Sars, to whom we owe so large a part of what is at present known respecting this curious subject of alternate generations.

Bougainvillia. (Bougainvillia superciliaris AG.)

The Bougainvillia (Fig. 92), is one of our most common Jelly-fishes, frequenting our wharves as well as our sea-shore during the spring. The tentacles are arranged in four bunches or clusters at the junction of the radiating tubes with the circular tube, from which they may be seen extending in every direction whenever these animals remain quietly suspended in the water,--a favorite attitude with them, and one which they retain sometimes for days, seeming to make no effort beyond that of gently playing their tentacles to and fro (Fig. 92). These tentacles are capable of immense extension, sometimes to ten or fifteen times the diameter of the bell. The proboscis is not simple as in the Sarsia, but looks like a yellow urn suspended at its four corners from the chymiferous tubes. The oral opening is entirely concealed by clusters of shorter tentacles surrounding the mouth in a close wreath, on which the eggs are supported. A highly magnified branch of the Hydroid stock from which this Medusa arises is represented in Fig. 93. There we see the little Jelly-fishes in different degrees of development on the stem, while in Figs. 94-97 they are given separately and still more enlarged. In Fig. 94 the outline of the Jelly-fish is still oval, the proboscis is but just formed, and the tentacles appear only as round swellings or knobs. In Fig. 95 a depression has taken place at the upper end, presently to be an opening, the proboscis is enlarged, and the tentacles lengthened, but still turned inward. In Fig. 96 the appendages of the proboscis are quite conspicuous, the tentacles are turned outward, and the Jelly-fish is almost ready to break from its attachment, having assumed its ultimate outline. Fig. 97 represents it just after it has separated from the stem, when it has only two tentacles at each cluster and simple knobs around the mouth, instead of the complicated branching tentacles of the adult.

Tubularia. (Tubularia Couthouyi AG.)

There are several other Tubularians common in our waters which should not be passed over without mention, although as this little book is by no means intended as a complete text-book, but rather as a volume of hints for amateur collectors, we would avoid as much as possible encumbering it with many names, or with descriptions already given in more comprehensive works. This Tubularia is interesting, however, from the fact that the Medusae buds are never freed from the stem, and do not develop into full-grown Jelly-fishes, but always remain abortive. Fig. 98 represents one head of such a Hydroid with the Medusae buds pendent from it in a thick cluster, while in Fig. 99 we have a few of them sufficiently magnified to show that, though presenting the four chymiferous tubes, they are otherwise exceedingly simple in structure, as compared with the free Jelly-fishes.

Hydractinia. (Hydractinia polyclina AG.)

This is another Tubularian, covering the surface of rocks in tide-pools, or attaching itself upon shells inhabited by hermit crabs. Indeed it was upon these shells that the Hydractinia was first noticed, and it was long supposed that the wanderings to which the little colony was thus subjected were necessary for its healthy development. But subsequent observations have shown that it attaches itself quite as frequently to the solid rock as to these nomadic shells. It has a rosy color, and, being very small, it looks, until one examines it closely, more like a thick red carpet of soft moss, than like a colony of animals. These communities are distinct in sex, the fertile individuals in each being either all male or all female. In Fig. 100 we have a portion of a female colony, representing one fertile head, in which the buds are crowded with Medusae; one sterile head, surrounded by its wreath of tentacles; and still another member of the society whose office is not fully understood, unless it be that of a kind of purveyor, catching food for the rest. Fig. 101 represents the corresponding individuals taken from a male colony. The sex makes little difference in the appearance of the reproductive heads. All the individuals of a Hydractinia colony are connected at the base by a horny network, rising occasionally into points of a conical or cylindrical shape. This polymorphism among the Tubularians is another evidence of the relation between the Siphonophorae, or floating Hydroids, and the fixed Hydroids.

Hybocodon. (Hybocodon prolifer AG.)

Among our Medusae derived from a Tubularian stock is the Hybocodon, viz. the hunchbacked Medusa (Fig. 102), a singular little Jelly-fish, odd and unsymmetrical in shape, as its name indicates, and interesting from its relations to one of our floating communities, the Nanomia, presently to be described. Instead of the evenly proportioned bell of the ordinary Medusae, the Hybocodon has a one-sided outline (Fig. 102), one large tentacle only being fully developed, while the others remain always abortive, so that the whole weight of the structure is thrown on one half of the bell. Upon this large tentacle small Jelly-fishes, similar to the original, are produced by budding, this process going on till ten or twelve such Jelly-fishes (Fig. 103) may be seen suspended from the tentacle. Up to this time it has remained connected with the Hydroid from which it arises, a rather large Tubularian, usually growing singly (Fig. 104), and of a deep orange-red in color. But at this stage of its existence it frees itself, and leads an independent life hereafter, swimming about with a quick, darting motion. In the account of the Nanomia, the homology between its scale, or abortive Medusa, and the Hybocodon, is traced in detail, and I need only allude to it here. Though this Medusa is so peculiar in appearance, the Tubularian from which it is derived is very like the Tubularia Couthouyi, already described. This is one of the instances before alluded to, in which closely allied forms give rise to very dissimilar ones, or, as in many cases, the very reverse of this takes place, and closely allied forms arise from very dissimilar ones.

Dysmorphosa. (Dysmorphosa fulgurans A. AG.)

Besides the budding at the base of the tentacle, as in Hybocodon, we find another mode of development among Hydroid Medusae, viz. that of budding from the proboscis. One of our most common little Jelly-fishes, the

Dysmorphosa (Fig. 105), to which we owe the occasional blue phosphorescence of the sea, so brilliant at times, buds in this manner. Fig. 105 represents an adult Dysmorphosa, on the proboscis of which may be seen three small buds in different stages of development. In Fig. 106 the proboscis is more enlarged, showing one of the little Jelly-fishes similar to the parent, just ready to drop off. We need not wonder at the immense number of these animals, with which the sea actually swarms at times, when we know that as fast as they are dropped, and it takes but a few days to complete their development, they each begin the same process; so that in the course of a week or ten days one such Medusa, supposing it to have produced six buds only, will have given rise to forty-two Jelly-fishes, thirty-six of which may be equally prolific in the same short period. These Medusae budding thus, and swimming about, carrying their young with them, bear such a close resemblance to the floating communities of Hydroids formerly known as Siphonophorae, that did we not know that some of them arise from Tubularians, it would be natural to associate them with the Siphonophorae.

Nanomia. (Nanomia cara A. AG.)

The Nanomia (Fig. 115), our free floating Hydroid, consists, when first formed, of a single Hydra containing an oblong oil bubble (Fig. 107). The whole organisation of such a Hydra is limited to a simple digestive cavity; it has, in fact, but one organ, and one function, and consists of an alimentary sac resembling the proboscis of a Medusa (Fig. 107); the oil bubble is separated from it by a transverse partition, and has no connection with the cavity. Presently, between the oil bubble and the cavity arise a number of buds of various character (Fig. 108), which we will describe one by one, beginning with those nearest the oil bubble, since these upper members of the little swimming community bear a very important part in its history. The infant community (Fig. 108) passes rapidly into the stage represented in Fig. 109, and then through all the stages intermediate between this and the adult, shown in its natural size in Fig. 115. The upper buds enlarge gradually, and soon take upon themselves a perfect Medusa structure (Fig. 110), with the exception of the proboscis, the absence of which is easily understood, when we find that these Medusae, serve the purpose of locomotion only, having no share in the function of feeding the community, so that a digestive apparatus would be quite superfluous for them. In every other respect they are perfect Medusae, attached to the Hydra as the Medusa buds always are when first formed, having the (four) chymiferous tubes, characteristic of all Hydroid Medusae, radiating from the centre to the periphery; two of these tubes are very winding, as may be seen in Fig. 110, while the other pair are straight. The Medusae themselves are heart-shaped in form, depressed at the centre of the upper surface, and bulging on either side into wing-like expansions, where they join the stem. These expansions interlock with one another, crossing nearly at right angles. The Medusae-like buds are the swimming bells; by their contractions, alternately taking in and throwing out the water, they impel the whole community forward, so that it seems rather to move like one animal, than like a combination of individuals.

Besides these locomotive members, the community contains three kinds of Hydrae arising as buds from the primitive Hydra below the swimming bells, the latter remaining always nearest the oil bubble at the top, while the first Hydra, the founder of the community, in proportion as the new individuals are added, is gradually pushed downward, and remains always at the end of the string, the stem of which is formed by the elongated neck of the primitive Hydra. All the three sets of Hydrae have certain features in common, while they have other distinguishing characteristics marking them as distinct individuals. They are all accompanied by triangular shields (Fig. 111), arising with them at the same point on the parent stem, and all are furnished with tentacles hanging down from the summit of the Hydra at the side opposite the shield. These facts are important to remember, since we shall presently perceive, upon analyzing their parts, that these Hydrae have a close homology to the Hybocodon. The tentacles differ in structure as well as in number for each kind of Hydra. Having shown in what characters they agree, let us now take each set individually, and see what differences they present.

In the first set which we will examine the Hydra is open-mouthed. Like the original Hydra, it is only a digestive tube, similar in all respects to the proboscis of a Medusa-disk. Its only function is that of feeding, and it shows a laudable fidelity to its calling, being very constantly and earnestly engaged in the work. Let us add, however, that in this instance the occupation is not a wholly selfish one, since the cavity of every Hydra communicates with that of the stem, and the food taken in at these over-gaping mouths, is at once circulated through all parts of the community, with the exception of the oil bubble, from which it is excluded by the transverse partition dividing it from all the lower members of the stock. The shields share in this general nourishment of the compound body by means of chymiferous tubes extending toward the outer surface, and opening into the cavity of the stem. The mouth of this Hydra is very flexible (Fig. 111), expanding and contracting at the will of the animal, and sometimes acting as a sucker, fastening itself, leech-like, on the object from which it seeks to draw its sustenance. (See Fig. 111.) The tentacles attached to this set of Hydrae are exceedingly long and delicate. They arise in a cluster at the upper and inner edge of the Hydra, just at its point of juncture with the stem, and being extremely flexible and contractile, their long tendril-like sprays are thrown out in an endless variety of attitudes. (See Fig. 115.) Along the whole length of this kind of tentacle are attached little pendent knobs at even distances; Fig. 112 represents such a knob greatly magnified, and absolutely paved with lasso-cells, the inner and smaller ones being surrounded by a row of larger ones.

In the third and last set of Hydrae (Fig. 114), the mouth is closed; they have, therefore, no share in feeding the community, but receive their nourishment from the cavity of the stem into which they open. They differ also from the others in having a single tentacle instead of a cluster, and on this tentacle the lasso-cells are scattered at uneven distances (Fig. 114). The special function of these closed Hydrae is yet to be explained; they have oil bubbles at their upper end (see Fig. 111, the top Hydra), and though we have never seen them drop off, it seems natural to suppose that they do separate from the parent stock, and found new communities similar to those from which they arise.

The intricate story of this singular compound existence does not end here. There is still another set of individuals whose share in maintaining the life of the community is by no means the least important. Little bunches of buds, of a different character from any described above, may be seen at certain distances along the lower part of the stem. These are the reproductive individuals. They are clusters of imperfect sexual Medusae, resembling the rudimentary Medusae of Tubularia (Fig. 99), which are never freed from the parent stem, but discharge their contents at the breeding season. Like many other compound Hydroids, the sexes are never combined, in one of these communities; they are always either male or female, and as those with female buds have not yet been observed, we can only judge by inference of their probable character. Front what is already known, however, of Hydroid communities of a like description, we suppose that the process of reproduction must be the same in these, and that the female stocks of Nanomia give birth to small Jelly-fishes, the eggs of which become oil bubbles, similar to that with which our little community began. (Fig. 108.)

By the time all these individuals have been added along the length of the stem, the stem itself has grown to be about three inches long (Fig. 115), though the tentacles hanging from the various members of the community give to the whole an appearance of much greater length. The motion of this little string of living beings is most graceful. The oil bubble (Fig. 116) at the upper end is their float; the swimming bells immediately below it (Fig. 110), by the convulsive contractions of which they move along, are their oars. The water is not taken in and expelled again by all the bells at once, but first from all the bells on one side, beginning at the lower one, and then from all those on the opposite side, beginning also at the lower one; this alternate action gives to their movements a swinging, swaying character, expressive of the utmost freedom and grace. Whether such a little community darts with a lightning-like speed through the water, or floats quietly up and down, for its movements are both rapid and gentle, it always sways in this way from side to side. Its beauty is increased by the spots of bright red scattered along the length of the stock at the base and tips of the Hydrae, as well as upon the tentacles. The movements and attitudes of the tentacles are most various. Sometimes they shoot them out in straight lines on either side, and then the aspect of the whole thing reminds one of a tiny chandelier in which the coral drops make the pendants, or they may be caught up in a succession of loops or floating in long streamers; indeed, there is no end to the fantastic forms they assume, ever astonishing you by some new combination of curves. The prevailing hue of the whole community is rosy, with the exception of the oil bubble or float, which looks a bright garnet color when seen in certain lights.

Let us now compare one of the Hydrae hanging from the stem (Fig. 113) with the Hybocodon (Fig. 102). The reader will remember the unsymmetrical bell of this singular Medusa, one half of its disk more largely developed than the other, with the proboscis hanging from the centre, and the cluster of tentacles from one side. Let us now split the bell so as to divide it in two halves with the proboscis hanging between them; next enlarge the side where there are no tentacles, and give it a triangular outline; then contract the opposite side so as to draw up the cluster of tentacles to meet the base of the proboscis, and what have we? The proboscis now corresponds to the

Hydra of our Nanomia, with the cluster of tentacles attached to its upper edge (Fig. 113), while the enlarged half of the bell represents the shield. If this homology be correct it shows that the Nanomia is not, as some naturalists have supposed all the Siphonophores to be, a single animal, its different parts being a mere collection of organs endowed with special functions, as feeding, locomotion, reproduction, &c., but that it is indeed a community of distinct individuals corresponding exactly to the polymorphous Hydroids, whose stocks are attached, such as Hydractinia, and differing from them only in being free and floating.

The homologies of the Siphonophorae or floating Hydroids, with many of the fixed Hydroids, is perhaps more striking when we compare the earlier stages of their growth. Suppose, for instance, that the planula of our Melicertum (See Fig. 81) should undergo its development without becoming attached to the ground,--what should we then have? A floating community (Fig. 83), including on the same stock like the Nanomia, both sterile and fertile Hydrae, from the latter of which Medusae bells are developed. The little Hydractinia community (Fig. 100), in which we have no less than four distinct kinds of individuals, each performing a definite distinct function, affords a still better comparison.

Physalia. (Physalia Arethusa TIL.)

Among the most beautiful of the Siphonophores, is the well-known Physalia or Portuguese man-of-war, represented in Fig. 117. The float above is a sort of crested sac or bladder, while the long streamers below consist of a number of individuals corresponding in their nature and functions to those composing a Hydroid community. Among them are the fertile and sterile Hydrae (Fig. 118), the feeders and Medusae bells (Fig. 119). The Physalia properly belongs to tropical waters, but sometimes floats northward, in the warm current of the Gulf Stream, and is stranded on Cape Cod. When found so far from their home, however, they have usually lost much of their vividness of color; to judge of their beauty one should see them in the Gulf of Mexico, sailing along with their brilliant float fully expanded, their crest raised, and their long tentacles trailing after them.

Another very beautiful floating Hydroid, occasionally caught in our waters, though its home is also far to the south, is the Velella (Fig. 120). It is bright blue in color, and in form not unlike a little flat boat with an upright sail. Its Medusa (Fig. 121) resembles so much that of some of our Tubularians, that it has actually been removed on this account from the old group of Siphonophorae, and placed next the Tubularians; another evidence of the close affinity between the former and the Hydroids.

MODE OF CATCHING JELLY-FISHES.

Not the least attractive feature in the study of these animals, is the mode of catching them. We will suppose it to be a warm, still morning at Nahant, in the last week of August, with a breath of autumn in the haze that softens the outlines of the opposite shore, and makes the horizon line a little dim. It is about eleven o'clock, for few of the Jelly-fishes are early risers; they like the warm sun, and at an earlier hour they are not to be found very near the surface. The sea is white and glassy, with a slight swell but no ripple, and seems almost motionless as we put off in a dory from the beach near Saunders's Ledge. We are provided with two buckets, one for the larger Jelly-fishes, the Zygodactyla, Aurelia, &c., the other for the smaller fry, such as the various kinds of Ctenophorae, the Tima, Melicertum, &c. Beside these, we have two nets and glass bowls, in which to take up the more fragile creatures that cannot bear rough handling. A bump or two on the stones before we are fairly launched, a shove of the oar to keep the boat well out from the rocks along which we skirt for a moment, and now we are off. We pull around the point to our left and turn toward the Ledge, filling our buckets as we go. Now we are crossing the shallows that make the channel between the inner and outer rocks of Saunders's Ledge. Look down,--how clear the water is and how lovely the sea-weeds, above which we are floating, dark brown and purple fronds of the Ulvae, and the long blades of the Laminaria with mossy green tufts between. As we issue from this narrow passage we must be on the watch, for the tide is rising, and may come laden with treasures, as it sweeps through it. A sudden cry from the oarsman at the bow, not of rocks or breakers ahead, but of "A new Jelly-fish astern!" The quick eye of the naturalist of the party pronounces it unknown to zoologists, un-described by any scientific pen. Now what excitement! "Out with the net!--we have passed him! he has gone down! no, there he is again! back us a bit." Here he is floating close by us; now he is within the circle of the net, but he is too delicate to be caught safely in that way, so, while one of us moves the net gently about, to keep him within the space enclosed by it, another slips the glass bowl under him, lifts it quickly, and there is a general exclamation of triumph and delight,--we have him. And now we look more closely; yes, decidedly he is a novelty as well as a beauty. (See Fig. 122, Ptychogena lactea A. Ag.) Those white mossy tufts for ovaries are unlike anything we have found before (Fig. 123), and not represented in any published figures of Jelly-fishes. We float about here for a while, hoping to find more of the same kind, but no others make their appearance, and we keep on our way to East Point, where there is a capital fishing ground for Medusae of all sorts. Here two currents meet, and the Jelly-fishes are stranded as it were along the line of juncture, able to move neither one way nor the other. At this spot the sea actually swarms with life; one cannot dip the net into the water without bringing up Pleurobrachia, Bolina, Idyia, Melicertum, &c., while the larger Zygodactyla and Aurelia float about the boat in numbers. These large Jelly-fishes produce a singular effect as one sees them at some depth beneath the water; the Aureliae, especially, with their large white disks, look like pale phantoms wandering about far below the surface; but they constantly float upward, and if not too far out of reach, one may bring them up by stirring the water under them with the end of the oar.

When we have passed an hour or so floating about just beyond East Point, and have nearly filled our buckets with Jelly-fishes of all sizes and descriptions, we turn and row homeward. The buckets look very pretty as they stand in the bottom of the boat with the sunshine lighting up their living contents. The Idyia glitters and sparkles with ever-changing hues, the Pleurobrachiae dart about, trailing their long graceful tentacles after them, the golden Melicerta are kept in constant motion by their quick, sudden contractions, and the delicate transparent Tima floats among them all, not the less beautiful because so colorless. There is an unfortunate Idyia, who, by some mistake, has got into the wrong bucket with the larger Jelly-fish, where a Zygodactyla has entangled it among his tentacles and is quietly breakfasting upon it.

During our row the tide has been rising, and as we near the channel of Saunders's Ledge, it is running through more strongly than before, and at the entrance of the shallows a pleasant surprise is prepared us; no less than half a dozen of our new friends (the Ptychogena as he has been baptized), come to look for their lost companion perhaps, await us there, and are presently added to our spoils. We reach the shore heavily laden with the fruits of our morning's excursion.

The most interesting part of the work for the naturalist is still to come. On our return to the Laboratory, the contents of the buckets are poured into separate glass bowls and jars; holding them up against the light, we can see which are our best and rarest specimens; these we dip out in glass cups and place by themselves. If any small specimens are swimming about at the bottom of the jar, and refuse to come within our reach, there is a very simple mode of catching them. Dip a glass tube into the water, keeping the upper end closed with your finger, and sink it till the lower end is just above the animal you want to entrap; then lift your finger, and as the air rushes out the water rushes in, bringing with it the little creature you are trying to catch. When the specimens are well assorted, the microscope is taken out, and the rest of the day is spent in studying the new Jelly-fishes, recording the results, making notes, drawings, &c.

Still more attractive than the rows by day are the night expeditions in search of Jelly-fishes. For this object we must choose a quiet night, for they will not come to the surface if the water is troubled. Nature has her culminating hours, and she brings us now and then a day or night on which she seems to have lavished all her treasures. It was on such a rare evening, at the close of the summer of 1862, that we rowed over the same course by Saunders's Ledge and East Point described above. The August moon was at her full, the sky was without a cloud, and we floated on a silver sea; pale streamers of the aurora quivered in the north, and notwithstanding the brilliancy of the moon, they too cast their faint reflection in the ocean. We rowed quietly along past the Ledge, past Castle Rock, the still surface of the water unbroken, except by the dip of the oars and the ripple of the boat, till we reached the line off East Point, where the Jelly-fishes are always most abundant, if they are to be found at all. Now dip the net into the water. What genie under the sea has wrought this wonderful change? Our dirty, torn old net is suddenly turned to a web of gold, and as we lift it from the water heavy rills of molten metal seem to flow down its sides and collect in a glowing mass at the bottom. The truth is, the Jelly-fishes, so sparkling and brilliant in the sunshine, have a still lovelier light of their own at night; they give out a greenish golden light as brilliant as that of the brightest glow-worm, and on a calm summer night, at the spawning season, when they come to the surface in swarms, if you do but dip your hand into the water it breaks into sparkling drops beneath your touch. There are no more beautiful phosphorescent animals in the sea than the Medusae; it would seem that the expression, "rills of molten metal" could hardly apply to anything so impalpable as a Jelly-fish, but, although so delicate in structure, their gelatinous disks give them a weight and substance; and at night, when their transparency is not perceived, and their whole mass is aglow with phosphorescent light, they truly have an appearance of solidity which is most striking, when they are lifted out of the water and flow down the sides of the net.

The various kinds present very different aspects; wherever the larger Aureliae and Zygodactylae float to the surface, they bring with them a dim spreading halo of light, the smaller Ctenophorae become little shining spheres, while a thousand lesser creatures add their tiny lamps to the illumination of the ocean; for this so-called phosphorescence of the sea is by no means due to the Jelly-fishes alone, but is also produced by many other animals, differing in the color as well as the intensity of their light, and it is a curious fact that they seem to take possession of the field by turns. You may row over the same course, which a few nights since glowed with a greenish golden light wherever the surface of the water was disturbed, and though equally brilliant, the phosphorescence has now a pure white light. On such an evening, be quite sure that when you empty your buckets on your return and examine their contents you will find that the larger part of your treasures are small crustacea (little shrimps). Of course there will be other phosphorescent creatures, Jelly-fishes, &c., among them, but the predominant color is given by these little crustacea. On another evening the light will have a bluish tint, and then the phosphorescence is principally due to the Dysmorphosa (Fig. 105).

Notwithstanding the beauty of a moonlight row, if you would see the phosphorescence to greatest advantage you must choose a dark night, when the motion of your boat sets the sea on fire around you, and a long undulating wave of light rolls off from your oar as you lift it from the water. On a brilliant evening this effect is lost in a great degree, and it is not until you dip your net fairly under the moonlit surface of the sea, that you are aware how full of life it is. Occasionally one is tempted out by the brilliancy of the phosphorescence, when the clouds are so thick that water, sky, and land become one indiscriminate mass of black, and the line of rocks can be discerned only by the vivid flash of greenish golden light, when the breakers dash against them. At such times there is something wild and weird in the whole scene, which at once fascinates and appalls the imagination; one seems to be rocking above a volcano, for the surface around is intensely black, except where fitful flashes or broad waves of light break from the water under the motion of the boat or the stroke of the oars. It was on a night like this, when the phosphorescence was unusually brilliant, and the sea as black as ink, the surf breaking heavily and girdling the rocky shore with a wall of fire, that our collector was so fortunate as to find in the rich harvest he brought home the entirely new and exceedingly pretty little floating Hydroid, described under the name of Nanomia (Fig. 115). It was in its very infancy (Fig. 108), a mere bubble, not yet possessed of the various appendages which eventually make up its complex structure; but it was nevertheless very important to have seen it in this early stage of its existence, since, when a few full-grown specimens were found in the autumn, which lived for some days in confinement and quietly allowed their portraits to be taken (see Fig. 115), it was easy to connect the adult animal with the younger phase of its own life and thus make a complete history.

Marine phosphorescence is no new topic, and we have dwelt too long, perhaps, upon a phenomenon that every voyager has seen, and many have described. Its effect is very different, when seen from the deck of a vessel, from its appearance as one floats through its midst, distinguishing the very creatures that produce it, and any account of the Medusae which did not include this most characteristic feature would be incomplete.

* * * * *

ECHINODERMS.

Our illustrations and descriptions of Echinoderms are scanty in comparison to those of the preceding class; for while, in consequence, perhaps, of the combined influence of the Gulf Stream and the cold arctic current on the New England shore, Acalephs are largely represented in our waters, our marine fauna is meagre in Echinoderms. But although we have few varieties, those which do establish themselves on a coast seemingly so ungenial for others of their kind, such as the Echinus, and our common Star-fish, for instance, thrive well and are very abundant. The class of Echinoderms includes five orders, viz. CRINOIDS, OPHIURANS, STAR-FISHES, SEA-URCHINS, and HOLOTHURIANS. The animals composing these orders differ so widely in appearance that it was very long before their true relations were detected, and it was seen that all their external differences were united under a common plan. Let us compare, for instance, the worm-like Holothurians (Figs. 124, 126, 127) with all the host of Star-fishes (Figs. 142, 146, 147) and Sea-urchins (Figs. 131, 139), or compare the radiating form of the Star-fish, its arms spreading in every direction, with the close spherical outline of the Sea-urchin, or the Crinoid floating at the end of a stem (Fig. 152) with either of these, and we shall cease to wonder that naturalists failed to find at once a unity of idea under all these varieties of execution. And yet the fundamental structure of the class of Echinoderms is represented as distinctly by any one of its five orders as by any other, and is absolutely identical in all. They differ only by trifling modifications of development.

In Echinoderms as a class, the body presents three regions differing in structure, and on the greater or less development of these regions or systems, as we may call them, their chief differences are based. Take, for instance, the dorsal system, the nature of which is explained by the name, indicating of course the back of the animal, though it does not necessarily imply the upper side of the body, since some of the Echinoderms, as the stemmed Crinoids, for example, carry the dorsal side downward, while the Star-fishes and Sea-urchins carry it upward, and the Holothurians, moving with the mouth forward, have the dorsal system at the opposite end of the body. Whatever the natural attitude of the animal, however, and the consequent position of the dorsal region, it exists alike in all the five orders, though it has not the same extent and importance in each. But in all it is made up of similar parts, bears the same relation to the rest of the body, has the same share in the general economy of the animal. And though when we compare the spreading back of a Star-fish with the small area on the top of a Sea-urchin, where all the zones unite, we may not at once see the correspondence between them, yet a careful comparison of all their structural details shows that they are both built with the same elements and represent the same region, though it is stretched to the utmost in the one case, and greatly contracted in the other.

This being true of the dorsal system, let us look at another equally important structural feature in this class. All Echinoderms have locomotive organs peculiar to themselves, a kind of suckers which may be more or less numerous, larger or smaller, in different species, but are always appendages of the same character. These are variously distributed over the body, but always with a certain regularity occupying definite spaces, shown by investigation to be homologous in all. For instance, the rays of the Star-fish correspond in every detail on their under side, along which the locomotive suckers run, with the zones on the Sea-urchin, from end to end of which the suckers are arranged; and the same is equally true of the distribution of the suckers on the Holothurians, Ophiurans, and Crinoids, though, as most persons are less familiar with these orders than with the other two, it might not be so easy to point out the coincidence to our readers. These suckers are called the ambulacra, the lines along which they run are called the ambulacral rows or zones, while the system of locomotion as a whole is known as the ambulacral system. Since these organs are thus regularly distributed over the body in distinct zones or rows, it follows that the latter must be divided by intervening spaces. These intervals are called the interambulacral spaces; but while in some orders they are occupied by larger plates and prominent spines, as in the Sea-urchin and Star-fish, in others they are either comparatively insignificant or completely suppressed, as in the Crinoids and Ophiurans. Such are the three regions or systems which by their greater or less development introduce an almost infinite variety of combinations into this highest class of Radiates. It may not be amiss before proceeding further to compare the five orders with reference to this point, and see which of these three systems has the preponderance in each one.

Taking the orders in their rank and beginning with the lowest, we find in the Crinoids that the dorsal system preponderates, being composed of highly complicated plates, and developed to such a degree as to form in many instances a stem by which the animal is attached to the ground, while the ambulacral system is limited to a comparatively small area, and the interambulacral system is wanting. The order of Crinoids has diminished so much in modern geological times that we must consult its fossil forms in order to understand fully the peculiar adaptation of the Echinoderm plan in this group.

In the Ophiurans, the dorsal system is still large, and though it no longer stretches out to form a stem, it folds over on the under side of the animal so as to enclose entirely the ambulacral system, forming a kind of shield for the arms. Here also the interambulacral system is wanting.

In the Star-fishes the dorsal system encroaches less upon the structure of the animal. The back and oral side here correspond exactly in size, and though the flat leathery upper surface of the animal, covered with spines, serves as a protection to the delicate ambulacral suckers which find their way between the rows of small plates along the under side of the arms, yet it does not enfold them as in the Ophiurans. On the contrary, in the Star-fishes the ambulacral rows are protected on either side by a row of the so-called interambulacral plates, through which no suckers pass.

In the Sea-urchin, the dorsal system is contracted to a minimum, forming a small area on the top of the animal, the rows of interambulacral plates which are separated and lie on either side of the ambulacra in the Star-fish being united in the Sea-urchin, and both the ambulacral and the interambulacral systems bent upward, meeting in the small dorsal area above, so as to form a spherical outline. Here the ambulacral and interambulacral systems have taken a great preponderance over the dorsal system, and the same is the case with the Holothurians, in which the same structure is greatly elongated, the dorsal system being thus pushed out as it were to the end of a cylinder, while the ambulacral and interambulacral systems run along its whole length. All Echinoderms without exception have ambulacral tubes, even though in some there are no external ambulacral suckers connected with them.

There is one organ peculiar to the class of Echinoderms, the general structure of which may be described here, since it is common to them all, with the exception of the Crinoids, the anatomy of which is, however, so imperfectly understood, that we are hardly justified in assuming that it does not exist even in that order. This organ is known as the madreporic body; it is a small sieve or limestone filter opening into a tube or canal; by means of this tube, which connects with the ambulacral system, the water from without, first filtered through the madreporic body and thus freed from any impurities, is conveyed to the ambulacra. In the more detailed account of the different orders we shall see what is the position of this singular organ in each group, and how it is adapted in them all to their special structure. The development of Echinoderms forms one of the most wonderful chapters in the annals of Natural History. Marvellous as is the embryonic history of the Acalephs, including all the different aspects they assume in the cycle of their growth, it is thrown into the shade by the transformations which Echinoderms undergo before assuming their adult condition. This singular mode of development, although it has features recalling the development of Jelly-fishes from

Hydroids, is nevertheless entirely distinct from it, and is known only in the class of Echinoderms. As the whole story is given at length in the chapter on the embryology of the Echinoderms, we need only allude to it here in general terms. We owe the discovery of this remarkable process to Johannes Mueller, one of the greatest anatomists of this century.

* * * * *

HOLOTHURIANS.

Synapta. (Synapta tenuis AYRES.)

This is one of the most curious of the Holothurians, and easily observed on account of its transparency, which allows us to see its internal structure. It has a long cylindrical body (Fig. 124) along the length of which run the five rows of ambulacra, which are in this instance closed tubes without any projecting suckers or locomotive organs of any kind attached to them, so that the name is retained only on account of their correspondence in position, and not from any similarity of function to the ambulacra in Star-fishes and Sea-urchins. But though the ambulacra in Synapta are in fact mere water-tubes like the vertical tubes in the Ctenophorae, by means of which the water, first filtered through the madreporic body, circulates along the skin, they are as organs perfectly homologous with the ambulacra in all other Echinoderms. The mouth has a circular tube around the aperture, and a wreath of branching tentacles encircling it. The habits of these animals are singular. They live in very coarse mud, but they surround themselves with a thin envelope of finer sand, which they form by selecting the small particles with their tentacles, and making a ring around their anterior extremity. This ring they then push down along the length of the body, and continue this process, adding ring after ring, till they have entirely encircled themselves with a sand tube. They move the rings down partly by means of contractions of the body, but also by the aid of innumerable appendages over the whole surface. To the naked eye these appendages appear like little specks on the skin; but under the microscope they are seen to be warts projecting from the surface, each one containing a little anchor with the arms turned upward (Fig. 125). Around the mouth these warts are larger, but do not contain any anchors. It will be seen hereafter that these appendages are homologous with certain organs in other Holothurians, the warts with the anchors corresponding to the limestone pavement covering or partially covering the surface of the Cuvieria, for instance, while those without anchors correspond to the so-called false ambulacra in Pentacta. By means of these appendages, though aided also by the contractions of the body, the Synaptae move through the mud and collect around themselves the sand tube in which they are encased. Their food is very coarse for animals so delicate in structure. When completely empty of food they are white, perfectly transparent, and the spiral tube forming the digestive cavity may be seen wound up and hanging loosely in the centre for the whole length of the body. In such a condition it is of a pale yellow color. But look at one that is gorged with food. The whole length of the alimentary canal is then crowded with sand, pebbles, and shells, distinctly seen through the transparent skin, and giving a dark gray color to the whole body. They swallow the sand for the sake of the nutritious substance it contains, and having assimilated and digested this, they then eject the harder materials. The motion of the body in consequence of its contractions is much like that of leeches, and on this account these Synaptae were long supposed to be a transition type between the Radiates and worms. The body grows to a great length, often half a yard and more, but constantly drops large portions from its posterior part, by means of its own contractions, or breaks itself up by the expulsion of the intestines, which are very readily cast out. The tentacles are hollow, consisting of a central rib with branches from either side. In the Synaptae, as in all the Holothurians, the madreporic body is placed near the mouth, between two of the ambulacra, and opposite the fifth or odd one. The tube, connecting with the central tube around the mouth, by means of which it communicates with the ambulacral tubes, is very short.

Caudina. (Caudina arenata STIMPS.)

Several other Holothurians are frequently met with on our shores. Among them is the Caudina arenata (Fig. 126), a small Holothurian, yellowish in color, and thick in texture, by no means so pretty as the white transparent Synapta; the tentacles are short, resembling a crown of cloves around the mouth. It lives in the sand, and may be found in great numbers on the sandy beaches after a storm.

Cuvieria. (Cuvieria squamata D. & K.)

The Holothurian of our coast, excelling all the rest in beauty, is the Cuvieria. (Fig. 127.) As it lies on the sand, a solid red lump, with neither grace of form nor beauty of color, even the vividness of its tint growing dull and dead when it is removed from its native element, certainly no one could suspect that it possessed any hidden charm; but place it in a glass bowl with fresh sea-water; the dull red changes to deep vivid crimson, the tentacles creep out (Fig. 127) softly, and slowly, till the mouth is surrounded by a spreading wreath, comparable for richness of tint, and for delicate tracery, to the most beautiful sea-weeds. These tentacles, when fully expanded, are as long as the body itself. A limestone pavement composed of numerous pieces covers almost the whole surface of the animal; this apparatus corresponds, as we have already mentioned, to the warts containing anchors in the Synapta; but in the latter, the limestone particles are smaller, whereas in the Cuvieria they are developed to a remarkable extent. This animal is very sluggish, the ambulacral suckers, found only on three of the tubes, being arranged in such a way as to form a sort of sole on which they creep; the sole is tough and leathery in texture, but free from the limestone pavement described above. The young (Figs. 128, 129) are very common, swimming freely about, and more readily found than the adult; they are of a bright vermilion color, but the tentacles hardly branch at that age, nor is the limestone pavement formed, which gives such a peculiar aspect to the full-grown animal. The young Cuvieria, somewhat older than that represented in Fig. 129, are found in plenty under stones at low-water mark, just after they have given up their nomadic habits, and when the limestone pavement begins to be developed.

Pentacta. (Pentacta frondosa JAeG.)

The highest of our Holothurians in structure, is the Pentacta. (Fig. 130) It is very rare on our beaches, though occasionally found under stones at low-water mark; farther north, in Maine, and at Grand Manan, it is very common, covering all the rocks near low-water mark. It is a chocolate brown in color, and measures, when fully expanded, some fifteen to eighteen inches in length. Unlike the Cuvieria, the ambulacral suckers are evenly distributed and almost equally developed on all the tubes; between the five rows of ambulacral suckers are scattered irregularly certain appendages resembling suckers, but found on examination not to be true locomotive suckers, and called on that account false ambulacra. These are the organs corresponding to the warts around the mouth of the Synapta. Although the ambulacral suckers are, as we have said, equally developed on all the tubes, yet the Pentacta does not use them indiscriminately as locomotive organs. In Pentacta, as well as in all Holothurians, whether provided with ambulacral suckers, or, like the Synapta and Caudina, deprived of them, the odd ambulacrum, viz. the one placed opposite the madreporic body, is always used to creep upon, and forms the under surface of the animal.

The correspondence between the different phases of growth in the young Pentacta, and the adult forms of the orders described above, the Synapta, Caudina, Cuvieria, and Pentacta itself, is a striking instance of the way in which embryonic forms illustrate the relative standing of adult animals. In the earlier stages of its development, the ambulacral tubes alone are developed in the Pentacta; in this condition it recalls the lower orders of Holothurians, as the Synapta and Caudina; then a sole is formed by the greater development of three of the ambulacra, and in this state it reminds us of the next in order, the Cuvieria, while it is only in assuming its adult form that the Pentacta develops its other ambulacra, with their many suckers.

The Pentacta resembles the Trepang, so highly valued by the Chinese as an article of food, and forms a not unsavory dish, having somewhat the flavor of lobster.

* * * * *

ECHINOIDS.

Sea-urchin. (Toxopneustes drobachiensis AG.)

Sea-urchins (Fig. 131) are found in rocky pools, hidden away usually in cracks and holes. They like to shelter themselves in secluded nooks, and, not satisfied even with the privacy of such a retreat, they cover themselves with sea-weed, drawing it down with their tentacles, and packing it snugly above them, as if to avoid observation. This habit makes them difficult to find, and it is only by parting the sea-weed, and prying into the most retired corners in such a pool, that one detects them. Their motions are slow, and they are less active than either the Star-fish or the Ophiuran, to both of which they are so closely allied.

Let us look at one first, as seen from above, with all its various organs fully extended. (Fig. 131.) The surface of the animal is divided by ten zones, like ribs on a melon, only that these zones differ in size, five broad zones alternating with five narrower ones. The broad zones, representing the interambulacral system, are composed of large plates, supporting a number of hard projecting spines, while the narrow zones, forming the ambulacral system, are pierced with small holes, arranged in regular rows, (Fig. 132,) through which extend the tentacles terminating with little cups or suckers. These zones converge towards the summit of the animal, meeting in the small area which here represents the dorsal system; this area is filled by ten plates, five larger ones at the extremity of the interambulacral zones, and five smaller ones at the extremity of the ambulacral zones. (Fig. 132.) In the five larger plates are the ovarian openings, so called because each one is pierced by a small hole through which the eggs are passed out, while in the five smaller plates are the eye-specks. The ovaries themselves consist of long pouches or sacs, carried along the inner side of each ambulacrum; one of these ovarian plates is larger than the others, and forms the madreporic body, being pierced with many minute holes; here, as in the Star-fish, it is placed between two of the ambulacral rows, and opposite the fifth or odd one. Looked at from the under or the oral side, as seen in Fig. 134, the animal presents the mouth, a circular aperture furnished with five teeth in its centre; these five teeth opening into a complicated intestine to be presently described. From the mouth, the ten zones diverge, curving upward to meet in the dorsal area on the summit of the body. (Fig. 133.)

Let us now examine the appearance and functions of the various appendages on the surface. The tentacles have a variety of functions to perform; they are the locomotive appendages, and for this reason, as we have seen, the zones along which they are placed are called the ambulacra, while the intervening spaces, or the broad zones, are called the interambulacra. It should not be supposed, however, that the locomotive appendages are the only ones to be found on the ambulacra, for spines occur on the narrow as well as on the broad zones, though the larger and more prominent ones are always placed on the latter. The tentacles are also subservient to circulation, for the water which is taken in at the madreporic body passes into all the tentacles, sometimes called on that account water-tubes. Beside these offices the tentacles are constantly busy catching any small prey, and conveying it to the mouth, or securing the bits of sea-weed with which, as has been said, these animals conceal themselves from observation. It is curious to see their fine transparent feelers, fastening themselves by means of the terminal suckers on such a floating piece of sea-weed, drawing it gently down and packing it delicately over the surface of the body. As locomotive appendages, the tentacles are chiefly serviceable on the lower or oral side of the animal, which always moves with the mouth downward. About this portion of the body the tentacles are numerous (Fig. 134) and large, and when the animal advances it stretches them in a given direction, fastens them by means of the suckers on some surface, be it of rock, or shell, or the side of the glass jar in which they are kept, and being thus anchored it drags itself forward. The tentacles are of a violet hue, though when stretched to their greatest length they lose their color, and become almost white and transparent; but in their ordinary condition the color is quite decided, and the rows along which they occur make as many violet lines upon the surface of the body.

Almost the sole function of the spines seems to be that of protecting the animal, and enabling it to resist the attacks of its enemies, the force of the waves, or any sudden violent contact with the rocks. The spines, when magnified, are seen to be finely ribbed for nearly the whole length (Fig. 135), the bare basal knob serving as the point of attachment for the powerful muscles, which move these spines on a regular ball-and-socket joint, the ball surmounting the tubercles (seen in Fig. 132), which fit exactly in a socket at the base of the spine. In a transverse section of a spine (Fig. 136), we see that the ribs visible on the outside are delicate columns placed closely side by side, and connected by transverse rods forming an exceedingly delicate pattern. Beside the tentacles and the spines, they have other external appendages, of which the function long remained a mystery, and is yet but partially explained; these are the so-called pedicellariae; they consist of a stem (s, Fig. 137), which becomes swollen (p, Fig. 137) into a thimble-shaped knob at the end (t, Fig. 137); this knob may seem solid and compact at first sight, but it is split into three wedges, which can be opened and shut at will. When open, these pedicellariae may best be compared to a three-pronged fork, except that the prongs are arranged concentrically instead of on one plane, and, when closed, they fit into one another as neatly as the pieces of a puzzle.

If we watch the Sea-urchin after he has been feeding, we shall learn, at least, one of the offices which this singular organ performs in the general economy of the animal. That part of his food which he ejects passes out at an opening on the summit of the body, in the small area where all the zones converge. The rejected particle is received on one of these little forks, which closes upon it like a forceps, and it is passed on from one to the other, down the side of the body, till it is dropped off into the water. Nothing is more curious and entertaining than to watch the neatness and accuracy with which this process is performed. One may see the rejected bits of food passing rapidly along the lines upon which these pedicellariae occur in greatest number, as if they were so many little roads for the conveying away of the refuse matters; nor do the forks cease from their labor till the surface of the animal is completely clean, and free from any foreign substance. Were it not for this apparatus the food thus rejected would be entangled among the tentacles and spines, and be stranded there till the motion of the water washed it away. These curious little organs may have some other office than this very laudable and useful one of scavenger, and this seems the more probable because they occur over the whole surface of the body, while they seem to pass the excrements only along certain given lines. They are especially numerous about the mouth, where they certainly cannot have this function; we shall see also that they bear an important part in the structure of the Star-fish, where there are no such avenues on the upper surface, for the passage of the refuse food, as occur on the Sea-urchin.

On opening a Sea-urchin, we find that the teeth (Fig. 138), which seem at first sight only like five little conical wedges around the mouth (Fig. 134), are connected with a complicated intestine, which extends spirally from the lower to the upper floor of the body, festooning itself from one ambulacral zone to the next, till it reaches the summit, where it opens. This intestine leads into the centre of the teeth, the jaws themselves, which sustain the teeth, being made up of a number of pieces, and moved by a complicated system of muscular bands. When the intestine is distended with food, it fills the greater part of the inner cavity; the remaining space is occupied in the breeding season by the genital organs. In a section of the Sea-urchin, one may also trace the tube by which the supply of water, first filtered through the madreporic body, is conveyed to the ambulacra; it extends from the summit of the body to the circular tube surrounding the mouth.

Echinarachnius. (Echinarachnius parma GRAY.)

Beside the Toxopneustes (Fig. 131) described above, we have another Sea-urchin very common along our shores. Among children who live near sandy beaches, they are well known as "sand-cakes" (Fig. 139), and indeed they are so flat and round, that, when dried and deprived of their bristles, they look not unlike a cake with a star-shaped figure on its surface. (Fig. 139.) When first taken from the water they are of a dark reddish brown color, and covered with small silky bristles. The disk is so flat, being but very slightly convex on the upper side, that one would certainly not associate it at first sight with the common spherical Sea-urchin or Sea-egg, as the Toxopneustes is sometimes called. But upon closer examination the delicate ambulacral tubes or suckers may be seen projecting from along the line of the ambulacra, as in the spherical Sea-urchin; and though these ambulacra become expanded near the summit into gill-like appendages, forming a sort of rosette in the centre of the disk, they are, nevertheless, the same organs, only somewhat more complicated. When such a disk is dried in the sun, and the bristles entirely removed, the lines of suture of the plates composing it, and corresponding exactly to those of the spherical Sea-urchin, may very readily be seen. (a and i, Fig. 139.)

This flat Sea-urchin or Echinarachnius, as it is called, belongs to a group of Sea-urchins known as Clypeastroids (shield-like Sea-urchins). In a section (Fig. 140) exposing the internal structure, one cannot but be reminded by its general aspect of an Aurelia. Could one solidify an Aurelia it would present much the same appearance; another evidence that all the Radiates are built on one plan, their differences being only so many modes of expressing the same structural idea. The teeth or jaws in this flat Sea-urchin are not so complicated as in the Toxopneustes, being simply flat pieces, arranged around the mouth (o, Fig. 140), without the apparatus of muscular bands by means of which the teeth are moved in the other genus. It is a curious fact, considered in relation to the general radiate structure of these animals, that the teeth, instead of moving up and down like the jaws in Vertebrates, or from right to left like those of Articulates, move concentrically, all converging towards the centre.

* * * * *

STAR-FISHES.

Star-fish. (Astracanthion berylinus AG.)

Although there is the closest homology of parts between the Star-fish and the Sea-urchin, the arrangement of these parts, and the external appearance of the animals, as a whole, are entirely different. The Star-fish has zones corresponding exactly to those of the Sea-urchin, but instead of being drawn together, and united at the summit of the animal, so as to form a spherical outline, they are spread out on one level in the shape of a star. This change in the general arrangement brings the eye-specks to the extremities of the arms, and places the ovarian openings in the angles between the arms. The madreporic body is situated on the upper surface of the disk (Fig. 142), at the angle between two of the arms, and consequently between two of the ambulacra, and opposite the odd one. The tube into which it opens, runs vertically from the upper floor of the disk to the lower, where it connects with the circular tube around the mouth, and thus communicates with all the ambulacral rows. The ambulacral zones which, in the Star-fish, have the shape of a furrow, run along the lower side of each ray (Fig. 141); the interambulacral zones are divided, their plates being arranged in rows along either side of the ambulacral furrows. The ambulacral furrow, like the ambulacral zone in the Sea-urchin, is pierced with numerous holes, alternating with each other in a kind of zigzag arrangement, one hole a little in advance, the next a little farther back, and so on, and through these holes pass the tentacles, terminating in suckers, as in the Sea-urchins, and serving as in them for locomotive organs. The most prominent and strongest spines are arranged upon the large interambulacral plates on both sides of the ambulacral furrows; but the upper surface of the animal is also completely studded with smaller spines, scattered at various distances, apparently without any regular arrangement. (Fig. 142.)

The position of the pedicellariae is quite different from that which they occupy in the Sea-urchin, where they are scattered singly between the spines and tentacles, though more regularly and closely grouped along the lines upon which the refuse food is moved off. In the Star-fish, on the contrary, these singular organs seem to be grouped for some special purpose around the spines, on the upper surface of the body. Every such spine swells near its point of attachment, thus forming a spreading base (Fig. 143), around which the pedicellariae are arranged in a close wreath, in the centre of which the summit of the spine projects; they differ also from those of the Sea-urchin in having two prongs instead of three. Other pedicellariae are scattered independently over the surface of the animal, but they are smaller than those forming the clusters and connected with the spines. The function of these organs in the Star-fish remains unexplained; the opening on the upper surface, through which the refuse food is thrown out, is in such a position that they evidently do not serve here the same purpose which renders them so useful to the Sea-urchin. Occasionally they may be seen to catch small prey with these forks, little Crustacea, for instance; but this is probably not their only office. The Star-fish has a fourth set of external appendages in the shape of little water-tubes. (Seen in Fig. 143.) The upper surface of the back consists of a strong limestone network (Fig. 144), and certain openings in this network are covered with a thin membrane through which these water-tubes project. It is supposed that water may be introduced into the body through these tubes; but while there can be no doubt that they are constantly filled with water, and are therefore directly connected with the circulation through the madreporic body (Fig. 145), no external opening has as yet been detected in them. The fact, however, that when these animals are taken out of their native element, the water pours out of them all over the surface of the back, so that they at once collapse and lose entirely their fulness of outline, seems to show that water does issue from those tubes. The ends of the arms are always slightly turned up, and at the summit of each is a red eye-speck. The tentacles about the eye become very delicate and are destitute of suckers.

These animals have singular mode of eating; they place themselves over whatever they mean to feed upon, as a cockle-shell for instance, the back gradually rising as they arch themselves above it; they then turn the digestive sac or stomach inside out, so as to enclose their prey completely, and proceed leisurely to suck out the animal from its shell. Cutting open any one of the arms we may see the yellow folds of the stomach pouches which extend into each ray; within the arms, extending along either side of the upper surface, are also seen the ovaries, like clusters of small yellow berries. Immediately below these, along the centre of the lower floor of each ray, runs the ridge formed by the ambulacral furrow, and upon either side of this ridge are placed the vesicles, by means of which the tentacles may be filled and emptied at the will of the animal; the rest of the cavity of the ray is filled by the liver. The mouth, which is surrounded by a circular tube, is not furnished with teeth, as in the Sea-urchin; but the end of each ambulacral ridge is hard, thus serving the purpose of teeth.

Cribrella. (Cribrella oculata FORBES.)

Our coast, as we have said, is not rich in the variety of Star-fishes. We have two large species, one of a dark-brown color (Fig. 132), the Astracanthion berylinus, and the other, the A. pallidus, of a pinkish tint; then there is the small Cribrella, inferior in structural rank to the two above mentioned. (Fig. 146.) This pretty little Star-fish presents the greatest variety of colors; some are dyed in Tyrian purple, others have a paler shade of the same hue, some are vermilion, others a bright orange or yellow. A glass dish filled with Cribrellae might vie with a tulip-bed in gayety and vividness of tints.

The disk of the Cribrella is smooth, instead of being covered, like the larger Star-fishes, with a variety of prominent appendages. The spines are exceedingly short, crowded like little warts over the surface. It is an interesting fact, illustrating again the correspondence between the adult forms of the lower orders and the phases of growth in the higher ones, that these spines have an embryonic character. One would naturally expect to find that these small spines of the adult Cribrella would differ from those of the other full-grown Star-fishes chiefly in size, that they would be a somewhat modified pattern of the same thing on a smaller scale; but when examined under the microscope, they resemble the spines of the higher orders in their embryonic condition; it is not, in fact, a difference in size merely, but a difference in degree of development. The Cribrella moves usually with two of the arms turned backward, and the three others advanced together, the two posterior ones being sometimes brought so close to each other as to touch for their whole length.

Hippasteria. (Hippasteria phrygiana AG.)

Beside these Star-fishes we have the pentagonal Hippasteria (Hippasteria phrygiana AG.), like a red star with rounded points, found chiefly in deep water, though it is occasionally thrown up on the beaches. It has but two rows of large tentacles, terminating in a powerful sucking disk. The pedicellariae on this Star-fish resemble large two-pronged clasps, arranged principally along the lower side. The pentagonal Star-fishes of our coast are in striking contrast to the long-armed species we have just described; they are edged with rows of large smooth plates, and do not possess the many prominent spines so characteristic of the ordinary Star-fishes.

Ctenodiscus. (Ctenodiscus crispatus D. & K.)

The Ctenodiscus (Ctenodiscus crispatus D. & K., Fig. 147), an inhabitant of more northern waters, but seeming also to be at home here occasionally, is another pentagonal Star-fish. It lives in deep water, and frequents muddy bottoms. The peculiar structure of their ambulacra has probably some reference to this mode of living, for they are entirely wanting in the sucking disks so characteristic of the other members of this class, and their tentacles are pointed, as if to enable them to work their way through the mud in which they make their home. The pointed tentacles of this genus are characteristic of a large group of Star-fishes, and it is an important fact, as showing their lower standing, that this feature, as well as the pentagonal outline, obtains in the earlier stages of growth of our more common Star-fishes, while in their adult condition they assume the deeply indented star-shaped outline, and have suckers at the extremities of the tentacles.

We find also among Star-fishes the same tendency to multiplication of parts so common among the Polyps and Acalephs. Our Solaster (Solaster endeca Forbes), for instance, has no less than twelve arms; it inhabits more northern latitudes, though sometimes found in our Bay; on the coast of Maine it is quite common, and occurs in company with another many-rayed species, the Crossaster papposa M. & T. The color of both of these Star-fishes is exceedingly varied; we find in the Solaster as many different hues as in the Cribrella, which it resembles in the structure of its spines, while in the Crossaster bands of different tints of red and purple are arranged concentrically, and the whole surface of the back is spotted with brilliantly-tinged tiny wreaths of water-tubes, crowded round the base of the different spines, which are somewhat similar to those of the Astracanthion.

OPHIURANS.

Ophiopholis. (Ophiopholis bellis LYM.)

There are but two species of the ordinary forms of Ophiurans in Massachusetts Bay; the white Amphiura (Amphiura squamata Sars), with long slender arms, and the spotted Ophiopholis (Fig. 148), with shorter and stouter arms, and in which the disk is less compact than in the Amphiura, and not so perfectly circular. All Ophiurans are difficult to find, from their exceeding shyness; they hide themselves in the darkest crevices, and though no eye-specks have yet been detected in them, they must have some quick perception of coming danger, for at the gentlest approach they instantly draw away and shelter themselves in their snug retreats. [Illustration: Fig. 149. One arm of Fig. 148; from the mouth side.]

They differ from the Star-fishes in having the disk entirely distinct from the arms; that is, the arms, instead of merging gradually into the disk, start at once from its margin. They have no interambulacral spaces or plates; but the whole upper surface is formed of large hard plates, which extend from the back over the sides of the arms to their lower surface, where they form a straight ridge along the centre. (Fig. 149.) The sides of these plates are pierced with holes, through which the tentacles pass; these have not, like those of the Star-fishes and Sea-urchins, a sucker at the extremity, but are covered with little warts or tubercles (Fig. 150); they are their locomotive appendages, and their way of moving is curious; they first extend one of the arms in the direction in which they mean to move, then bring forward two others to meet them, three arms being thus usually in advance, and then they drag the rest of the body on. They move with much more rapidity, and seem more active, than the Star-fishes; probably owing to the greater independence of the arms from the disk. The spines project along the margin of the arms, and not over the whole surface, the back of the arms being perfectly free from any appendages, and presenting only the surface of the plates. The madreporic body is formed by a plate on the lower side of the disk, in a position corresponding to that which it occupies in the young Star-fish; this plate is one of the large circular shields occupying the interambulacral spaces around the mouth. (Fig. 149.) On each side of the arms, where they join the disk, are slits opening into the ovarian pouches. They have no teeth; but the hard ridge at the oral end of the ambulacra, extending toward the mouth in Star-fish, is still more distinct and sharper in the Ophiurans, approaching more nearly the character of teeth.

Astrophyton. (Astrophyton Agassizii STIMP.)

A singular species of Ophiuran, known among fishermen as the "Basket-fish," (Fig. 151,) is to be found in Massachusetts Bay. Its arms are very long in comparison to the size of the disk, and divide into a vast number of branches. In moving, the animal lifts itself on the extreme end of these branches, standing as it were on tiptoe (Fig. 151), so that the ramifications of the arms form a kind of trellis-work all around it, reaching to the ground, while the disk forms a roof. In this living house with latticed walls small fishes and other animals are occasionally seen to take shelter; but woe to the little shrimp or fish who seeks a refuge there, if he be of such a size as to offer his host a tempting mouthful; he will fare as did the fly who accepted the invitation of the spider. These animals are exceedingly voracious, and sometimes, in their greediness for food, entangle themselves in fishing lines or nets. When disturbed, they coil their arms closely around the mouth, assuming at such times a kind of basket-shape, from which they derive their name.

This Basket-fish is honorably connected with our early colonial history, being thought worthy, by no less a personage than John Winthrop, Governor of Connecticut, who, as he says, "had never seen the like," to be sent with "other natural curiosities of these parts" to the Royal Society of London, in 1670. He accompanies the specimen with a minute description, omitting "other particulars, that we may reflect a little upon this elaborate piece of nature." His account is as graphic as it is accurate, and we can hardly give a better idea of the animal than by extracting some portions of it. "This Fish," he says, "spreads itself from a Pentagonal Root, which incompasseth the Mouth (being in the middle), into 5 main Limbs or branches, each of which, just at issuing out from the Body, subdivides itself into two, and each of these 10 branches do again divide into two parts, making 20 lesser branches; each of which again divide into two smaller branches, making in all 40. These again into 80, and these into 160; and these into 320; these into 640; into 1280; into 2560; into 5120; into 10,240; into 20,480; into 40,960; into 81,920; beyond which the further expanding of the Fish could not be certainly trac'd";--a statement which we readily believe, wondering only at the patience which followed this labyrinth so far.

In a later letter, after having had an interview with the fisherman who caught the specimen, and, as he says, "asked all the questions I could think needful concerning it," the Governor proceeds to tell us that it was caught "not far from the Shoals of Nantucket (which is an Island upon the Coast of New England)," and that when "first pull'd out of the water it was like a basket, and had gathered itself round like a Wicker-basket, having taken fast hold upon that bait on the hook which he" (the fisherman) "had sunk down to the bottom to catch other Fish, and having held that within the surrounding brachia would not let it go, though drawn up into the Vessel; until, by lying a while on the Deck, it felt the want of its natural Element; and then voluntarily it extended itself into the flat round form, in which it appear'd when present'd to your view." The Governor goes on to reflect in a philosophical vein upon the purpose involved in all this complicated machinery. "The only use," he says, "that could be discerned of all that curious composure wherewith nature had adorned it seems to be to make it as a purse-net to catch some other fish, or any other thing fit for its food, and as a basket of store to keep some of it for future supply, or as a receptacle to preserve and defend the young ones of the same kind from fish of prey; if not to feed on them also (which appears probable the one or the other), for that sometimes there were found pieces of Mackerel within that concave. And he, the Fisherman, told me that once he caught one, which had within the hollow of its embracements a very small fish of the same kind, together with some piece or pieces of another fish, which was judged to be of a Mackerel. And that small one ('tis like) was kept either for its preservation or for food to the greater; but, being alive, it seems most likely it was there lodged for safety, except it were accidentally drawn within the net, together with that piece of fish upon which it might be then feeding." The account concludes by saying, "This Fisherman could not tell me of any name it hath, and 'tis in all likelihood yet nameless, being not commonly known as other Fish are. But until a fitter English name be found for it, why may it not be called (in regard of what hath been before mentioned of it) a Basket-Fish, or a Net-Fish, or a Purs-net-Fish?" And so it remains to this day as the Governor of Connecticut first christened it, the Basket-fish.

* * * * *

CRINOIDS.

The Crinoids are very scantily represented in the present creation. They had their day in the earlier geological epochs, when for some time they remained the sole representatives of their class, and were then so numerous that the class of Echinoderms, with only one order, seemed as full and various as it now does with five. The different forms they assumed in the successive geological periods are particularly instructive; these older Crinoids combined characters which foreshadowed the advent of the Ophiurans, the true Star-fishes, and the Sea-urchins; and so prominently were their prophetic characters developed, that many of them are readily mistaken for Star-fishes or Sea-urchins.

In later times the group of Crinoids has been gradually dwindling in number and variety. Its present representatives are the Pentacrini of Porto Rico and the coast of Portugal, the lovely little Rhizocrinus of the Atlantic, dredged first by the younger Sars on the coast of Norway, attached throughout life to a stem, and the Comatula, which has a stem only in the early stages of its growth, but is free when adult. The Pentacrinus bears the closer relation to the more ancient Crinoids (Fig. 152), which were always supported on a stem, while it is only in more recent periods that we find the free Crinoids, corresponding to the Comatula.

Comatula. (Alecto meridionalis AG.)

One large species of Comatula (Alecto Eschrichtii M. & T.) is known on our coast, off the shores of Greenland, where it has been dredged at a depth of about one hundred and fifty fathoms, and young specimens of the same species have been found as far south as Eastport, Maine. The species selected for representation here, however, (Fig. 153,) is one quite abundant along the shores of South Carolina. It is introduced instead of the northern one, because the latter is so rare that it is not likely to fall into the hands of our readers. The annexed drawing (Fig. 154, magnified from Fig. 153) represents a group of the young of the Charleston Comatula, still attached to the parent body by their stems, and in various stages of development. At first sight, the Comatula, or, as it is sometimes called, the feather-star, resembles an Ophiuran; but on a closer examination we find that the arms are made up of short joints; and along the sides of the arms, attached to each joint, are appendages resembling somewhat the beards of a feather, and giving to each ray the appearance of a plume; hence the name of feather-star. On one side the arms are covered with a tough skin, through which project the ambulacrae, and on the same side of the disk are situated the mouth and the anus; the latter projects in a trumpet-shaped proboscis. On the opposite side of the disk the Comatula is covered with plates, arranged regularly around a central plate, which is itself covered with long cirri.

We are indebted to Thompson for the explanation of the true relations of the young Comatula to the present Pentacrinus and the fossil Crinoids. Supposing these young to be full-grown animals, he at first described them as living representatives of the genus Pentacrinus; it was only after he had watched their development, and ascertained by actual observation that they dropped from their stem, to lead an independent life as free Comatulae, that he fully understood their true connection with the past history of their kind, as well as with their contemporaries. In Fig. 153, a faint star-like dot (y) may be seen attached to the side of the disk by a slight line. In Fig. 154, we have that minute dot as it appears under the microscope, magnified many diameters; when it is seen to be a cirrus of a Comatula, with three small Pentacrinus-like animals growing upon it, in different stages of development. In the upper one, the branching arms and the disk, with its many plates, are already formed; and though in the figure the rays are folded together, they are free, and can be opened at will. In the larger of the two lower buds, the plates of the disk are less perfect, and the arms are straight and simple, without any ramifications, though they are free and movable, whereas, in the smaller one, they are folded within the closed bud.

* * * * *

EMBRYOLOGY OF ECHINODERMS.

All Radiates have a special mode of development, as distinct for each class as is their adult condition, and in none are the stages of growth more characteristic than in the Echinoderms. In the Polyps, the division of the body into chambers, so marked a feature of their ultimate structure, takes place early; in the Acalephs, the tubes which traverse the body are hollowed out of its mass in the first stages of the embryonic growth, and we shall see that in the Echinoderms also, the distinctive feature of their structure, viz. the enclosing of the organs by separate walls, early manifests itself. This peculiarity gives to the internal structure of these animals so individual a character, that some naturalists, overlooking the law of radiation, as prevalent in them as in any members of this division, have been inclined to separate them, as a primary division of the animal kingdom, from the Polyps and Acalephs, in both of which the body-wall furnishes the walls of the different internal cavities, either by folding inwardly in such a manner as to enclose them, as in the Polyps, or by the cavities themselves being hollowed out of the general mass, as in the Acalephs.

Star-fish. (Astracanthion.)

The egg of the Star-fish, when first formed, is a transparent, spherical body, enclosing the germinative vesicle and dot. (See Fig. 155.) As soon as these disappear, the segmentation of the yolk begins; it divides first into two portions (see Fig. 156), then into four, then into eight, and so on; but when there are no more than eight bodies of segmentation (see Fig. 157), they already show a disposition to arrange themselves in a hollow sphere, enclosing a space within, and by the time the segmentation is completed, they form a continuous spherical shell. At this time the egg, or, as we will henceforth call it, the embryo, escapes and swims freely about. (See Fig. 158.) The wall next begins to thin out on one side, while on the opposite side, which by comparison becomes somewhat bulging, a depression is formed (m a, Fig. 159), gradually elongating into a loop hanging down within the little animal, and forming a digestive cavity. (d, Fig. 160.) At this stage it much resembles a young Actinia. The loop spreads somewhat at its upper extremity, and at its lower end is an opening, which at this period of the animal's life serves a double purpose, that of mouth and anus also, for at this opening it both takes in and rejects its food. We shall see that before long a true mouth is formed, after which this first aperture takes its place opposite the mouth, retaining only the function of the anus. Presently from the upper bulging extremity of the digestive cavity, two lappets, or little pouches, project (w w' Fig. 161); they shortly become completely separated from it, and form two distinct hollow cavities (w w', Fig. 162). Here begins the true history of the young Star-fish, for these two cavities will develop into two water-tubes, on one of which the back of the Star-fish, that is, its upper surface, covered with spines, will be developed, while on the other, the lower surface, with the suckers and tentacles, will arise. At a very early stage one of these water-tubes (w', Fig. 163) connects with a smaller tube opening outwards, which is hereafter to be the madreporic body (b, Fig. 163). Almost until the end of its growth, these two surfaces, as we shall see, remain separate, and form an open angle with one another; it is only toward the end of the development that they unite, enclosing between them the internal organs, which have been built up in the mean while.

At about the same time with the development of these two pouches, so important in the animal's future history, the digestive cavity becomes slightly curved, bending its upper end sideways till it meets the outer wall, and forms a junction with it (m, Fig. 164). At this point, when the juncture takes place, an aperture is presently formed, which is the true mouth. The digestive sac, which has thus far served as the only internal cavity, now contracts at certain distances, and forms three distinct, though connected cavities, as in Fig. 163; viz. the oesophagus leading directly from the mouth (m) to the second cavity or stomach (d), which opens in its turn into the third cavity, the alimentary canal. Meanwhile the water-tubes have been elongating till they now surround the digestive cavity, extending on the other side of it beyond the mouth, where they unite, thus forming a Y-shaped tube, narrowing at one extremity, and dividing into two branches toward the other end. (Fig. 165.)

On the surface where the mouth is formed, and very near it on either side, two small arcs arise, as v in Fig. 162; these are cords consisting entirely of vibratile cilia. They are the locomotive organs of the young embryo, and they gradually extend until they respectively enclose nearly the whole of the upper and lower half of the body, forming two large shields or plastrons. (Figs. 165, 166.) The corners of these shields project, slightly at first (Fig. 165), but elongating more and more until a number of arms are formed, stretching in various directions (Figs. 166, 167), and, by their constant upward and downward play, moving the embryo about in the water.

At this stage of the growth of the embryo, we have what seems quite a complicated structure, and might be taken for a complete animal; this is after all but the prelude to its true Star-fish existence. While these various appendages of the embryo have been forming, changes of another kind have taken place; on one of the two water-tubes above mentioned (w'), at the end nearest the digestive cavity, a number of lobes are formed (t, Fig. 166); this is the first appearance of the tentacles. In the same region of the opposite water-tube (w) a number of little limestone rods arise, which eventually unite to form a continuous network; this is the beginning of the back of the Star-fish (r, Fig. 166), from which the spines will presently project. When this process is complete, the whole embryo, with the exception of the part where the young Star-fish is placed, grows opaque; it fades, as it were, begins to shrink and contract, and presently drops to the bottom, where it attaches itself by means of short arms (f f', Fig. 166), covered with warts, which act as suckers, and are placed just above the mouth. As soon as the Star-fish has thus secured itself, it begins to resorb the whole external structure described above; the water-tubes, the plastrons, and the complicated system of arms connected with them, disappear within the little Star-fish; it swallows up, so to speak, the first stage of its own existence; it devours its own larva, which now becomes part and parcel of the new animal. Next the two surfaces, the back and lower surface, on which the arms are now marked out, while the tentacles, suckers, and spines have already assumed a certain prominence, approach each other. At this time, however, the arms are not in one plane; both the back and the lower surface are curved in a kind of spiral; they begin to flatten; the arms spread out on one level,--and now the two surfaces draw together, meeting at the circumference, and enclosing between them the internal organs, which, as we have seen, are already formed and surrounded by walls of their own, before the two walls of the body, close thus over them. Fig. 168 represents the upper surface of the Star-fish just before this junction takes place. The complicated structure of the Brachiolaria, as the larva of the Star-fish has been called, hitherto so essential to the life of the animal, by which it has been supported, moved about in the water, and provided with food during its immature condition, has made a final contribution to its further development by the process of resorption described above, and has wholly disappeared within the Star-fish. At this stage the rays are only just marked out, as five lobes around the margin; Fig. 169 represents the lower surface at the same moment, with the open mouth (m), around which the tentacles (t) are just beginning to appear; while Fig. 170 shows us the animal at a more advanced stage, after the two surfaces have united. It has now somewhat the outline of a Maltese cross, the five arms being more distinctly marked out, while the tentacles have already attained a considerable length (Fig. 171), and the dorsal plates have become quite distinct. Fig. 172 represents the same animal, at the same age, in profile. This period, in which we have compared the form of the Star-fish to that of a Maltese cross, is one of long duration; two or three years must elapse before the arms will elongate sufficiently to give it a star-shaped form, and before the pedicellariae make their appearance, and it is only then that it can be at once recognized as the young of our common Star-fish. Even then, after it has assumed its ultimate outline, it lacks some features of the adult, having only two rows of tentacles, whereas the full-grown Star-fish has four.

This extraordinary process of development which we have analyzed thus at length in the history of the Star-fish, but which is equally true of all

Echinoderms, has been hitherto described (so far as it was known) under the name of the plutean stages of growth. In these early stages the young, or the so called larvae of Echinoderms, have received the name of Pluteus on account of their ever-changing forms. Let us look for a moment at the plutean stages of the Sea-urchin, as they differ in some points from those of the Star-fish. In the Pluteus of our common Sea-urchins (see Fig. 176), the arms are supported by a framework of solid limestone rods, which do not exist in that of the Star-fish, and which give to the larva of the Sea-urchin a remarkable rigidity. They are formed very early, as may be seen in Fig. 173, representing the little Sea-urchin before any arms are discernible, though the limestone rods are quite distinct. Figs. 173, 174, 175, may be compared with Figs. 160, 162, 165, of the young Star-fish, where it will be seen that the general outline is very similar, though, on account of the limestone rods, the Pluteus of the Sea-urchin seems somewhat more complicated. In Fig. 176 the young Sea-urchin has so far encroached upon the Pluteus that it forms the essential part of the body, the arms and rods appearing as mere appendages. Fig. 177 shows the same animal when we looked down upon it in its natural attitude; the Sea-urchin is carried downward, and the arms stretch in every direction around it. In Fig. 178 the Plutens is already in process of absorption; in Fig. 179 it has wholly disappeared; in Figs. 180 and 181 we have different stages of the little Sea-urchin, with its spines and suckers of a large size and in full activity. The appearance of the Sea-urchin, as soon as this larva or Pluteus is completely absorbed, is much more like that of the adult than is the Star-fish at the same stages, in which, as we have seen, there is a transition period of considerable duration.

Ophiurans.

Fig. 183 represents an Ophiuran undergoing the same process of growth, at a period when the larva is most fully developed, and before it begins to fail. By the limestone rods which support the arms, the Pluteus of the Ophiuran, here represented, resembles that of the Sea-urchin more than that of the Star-Fish, while by the character of the water-tubes and by its internal organization it is more closely allied to the latter. It differs from both, however, in the immense length of two of the arms; these arms being the last signs of its plutean condition to disappear; when the young Ophiuran has absorbed almost the whole Pluteus, it still goes wandering about with these two immense appendages, which finally share the fate of all the rest. Fig. 182

represents an Ophiuran at the moment when the process of resorption is nearly completed, though the arms of the Pluteus, greatly diminished, are still to be seen protruding from the surface of the animal.

This mode of development, though common to all Echinoderms, appears under very different conditions in some of them. There are certain Star-fishes, Ophiurans, and Holothurians, passing through their development under what is known as the sedentary process. The eggs are not laid, as in the cases described above, but are carried in a sort of pouch over the mouth of the parent animal, where they remain till they attain a stage corresponding to that of Fig. 168 of the Star-fish, and having much the same cross-shaped outline, when they escape from the pouch (as the young Ophiopholis, Fig. 184), and swim about for the first time as free animals. Fig. 185 represents a cluster of young Star-fishes of the sedentary kind at about this period. But while this mode of growth seems at first sight so different, we shall find, if we look a little closer, that it is essentially the same, and that, though the circumstances under which the development takes place are changed, the process does not differ. The little Star-fish or Ophiuran, in the pouch, becomes surrounded by the same plutean structure as those which are laid in the egg; it is only more contracted to suit the narrower space in which they have to move; and the water-tubes on which the upper and lower surfaces of the body arise, the shields, spreading out into arms at the corners, exist, fully developed or rudimentary, in the one as much as in the other, and when no longer necessary to its external existence they are resorbed in the same way in both cases. This singular process of development has no parallel in the animal kingdom, although the growth of the young Echinoderm on the Brachiolaria may at first sight remind us of the budding of the little Medusa on the Hydroid stock, or even of the passage of the insect larva into the chrysalis. But in both these instances, the different phases of the development are entirely distinct; the Hydroid stock is permanent, continuing to live and grow and perform its share in the cycle of existence to which it belongs, after the Medusa has parted from it to lead a separate life, or if the latter remains attached to the parent stock, after it has entered upon its own proper functions. The life of the caterpillar, chrysalis and butterfly, is also distinct and definitely marked; the moment when the animal passes from one into the other cannot be mistaken, although the different phases are carried on successively and not simultaneously, as in the case of the Acalephs. But in the Echinoderms, on the contrary, though the aspect of the Brachiolaria, or plutean stage, is so different from that of the adult form, that no one would suppose them to belong to the same animal, yet these two stages of growth pass so gradually into one another, that one cannot say when the life of the larva ceases, and that of the Echinoderm begins.

The bearing of embryology upon classification is becoming every day more important, rendering the processes of development among animals one of the most interesting and instructive studies to which the naturalist can devote himself, in the present state of his science. The accuracy of this test, not only as explaining the relations between animals now living, but as giving the clew to their connection with those of past times, cannot but astonish any one who makes it the basis of his investigations. The comparison of embryo forms with fossil types is of course difficult, and must in many instances be incomplete, for while, in the one case, death and decay have often half destroyed the specimen, in the other, life has scarcely stamped itself in legible characters on the new being. Yet, whenever such comparisons have been successfully carried out, the result is always the same; the present representatives of the fossil types recall in their embryonic condition the ancient forms, and often explain their true position in the animal kingdom. One of the most remarkable examples of this in the type we are now considering, is that of the Comatula already mentioned. Its condition in the earlier stages of growth, when it is provided with a stem, at once shows its relation to the old stemmed Crinoids, the earliest representatives of the class of Echinoderms.

These coincidences are still more striking among living animals, where they can be more readily and fully traced, and often give us a key to their relative standing, which our knowledge of their anatomical structure fails to furnish. This is perhaps nowhere more distinctly seen than in the type of Radiates, where the Acalephs in their first stages of growth, that is, in their Hydroid condition, remind us of the adult forms among Polyps, showing the structural rank of the Acalephs to be the highest, since they pass beyond a stage which is permanent with the Polyps; while the adult forms of the Acalephs have in their turn a certain resemblance to the embryonic phases of the class next above them, the Echinoderms. Within the limits of the classes, the same correspondence exists as between the different orders; the embryonic forms of the higher Polyps recall the adult forms of the lower ones, and the same is true of the Acalephs as far as these phenomena have been followed and compared among them. In the class of Echinoderms the comparison has been carried out to a considerable extent, their classification has hitherto been based chiefly upon the ambulacral system, so characteristic of the class, but so unequally developed in the different orders. This places the Holothurians, in which the ambulacral system has its greatest development, at the head of the class; next to them come the Sea-urchins or Echinoids; then the Star-fishes; then the Ophiurans and Crinoids, in which the ambulacral system is reduced to a minimum. Another basis for classification in this type, which gives the same result, is the indication of a bilateral symmetry in some of the orders. In the Holothurians, for instance, there is a decided tendency toward the establishment of a posterior and anterior extremity, of a right and left, an upper and lower side of the body. In the Sea-urchins, in many of which the mouth is out of centre, placed nearer one side than the other, this tendency is still apparent, while in the three lower groups, the Star-fishes, Ophiurans, and Crinoids, it is almost entirely lost, in the equal division of identical parts radiating from a common centre. A comparison of the embryonic and adult forms in these orders, confirms entirely this classification based upon structural features. The Star-fishes, in their earlier stages, resemble the mature Ophiurans, while the Crinoids, the lowest group of all, retain throughout their whole existence many features characteristic of the embryonic conditions of the higher Echinoderms. In this principle of classification, already so fertile in results, we may hope to find, in some instances, the solution of many perplexing points respecting the structural rank of animals, the confirmation of classifications already established; in others, an insight into the true relations of groups which have hitherto been divided upon purely arbitrary grounds.

* * * * *

DISTRIBUTION OF LIFE IN THE OCEAN.

We have seen that while our bay is rich in certain species, it is wholly deficient or but scantily supplied with others, and that the character of the animals inhabiting its waters is more or less directly connected with general physical conditions. Such an area, limited though it be, gives us some insight into the laws which, in their wider application, control the distribution of marine life along the shores of the most extensive continents. The coast of Massachusetts, taken as a whole, is like that of New England generally, a rocky coast; yet it has its sandy and muddy beaches, and though it lies for a great part open to the sea, it has nevertheless its sheltered harbors, its quiet bays and snug recesses.

A comparison of these limited localities with far more extensive reaches of shore, where similar physical conditions prevail, shows that they reproduce, in fainter and less various characters of course, in proportion to their narrower boundaries, but still with a certain fidelity, the same combinations of animal and vegetable life. In other words, a sandy beach, however small, gives us some idea of the nature of the animals we may look for on any sandy coast, as, for instance, clams of various kinds, razor-shells, quahogs, snails, &c., creatures who can penetrate the sand, drag themselves through it or over it, leaving their winding trails as they go, and to whom the conditions prevailing in such spots are genial. So the narrowest mud flat on the sea-shore or muddy beach will give us the same dead and inanimate aspect which characterizes a more extensive coast of like character, where the gases always generated in mud are deadly to many kinds of animals. The beings who find a home in such localities are of closely allied species, chiefly a variety of worms, who burrow their way into the mud, and seem to court the miasma so fatal to other creatures. The same is true of any stony beach or rocky shore not more than a quarter of a mile in length; it gives us an idea of the animal population on any similar coast of greater extent.

These correspondences are of course modified by differences in climatic conditions. The animals on a sandy beach or a rocky shore, on the coast of Great Britain, for instance, are not absolutely identical with those of a sandy beach or a rocky shore on the coast of New England, but they are more or less nearly related to them. Naturalists refer to this reiteration, all the world over, of like organic combinations under similar circumstances, when they speak of "representative species." The aggregate result is the same, though the individual forms are slightly modified. And here lies one secret of the infinite variety in nature, by which the old seems ever new, and the same thought has an eternal freshness and originality, endlessly repeated, yet never hackneyed.

In this sense our bay presents, on a miniature scale, a variety of physical and organic combinations, which may be compared to those more extensive divisions in the geographical distribution of animals and plants, called by naturalists zoological or botanical provinces or districts, the animal and vegetable populations of which are technically designated as their faunae and florae. Such organic realms, as we may call them, have long been recognized on land, and the most extensive among them are easily distinguished. No one will fail to recognize the tropical zone, with its royal dynasty of palms and all the accompanying glories of a tropical vegetation, its birds of brilliant plumage, its large Mammalia, lions, tigers, panthers, elephants, and its great rivers haunted by gigantic reptiles. Nor is the representation of vegetable and animal life less characteristic in the temperate zone, where the oak is monarch of the woods, with all his attendant court of elms, walnuts, beeches, birches, maples, and the like, where birds of more sober hues, but sweeter voices, take the place of the brilliant parrots and many-tinted humming-birds of the tropical forest; while buffaloes, bears, wolves, foxes, and deer represent the larger Mammalia. In the arctic zone, though marked by peculiar and distinctive features, vegetation has dwindled to a minimum; the birds are chiefly gulls and ducks, which go there for the breeding season in the summer, and the reindeer and polar bears are almost sole possessors of the snow and ice-fields; but this meagreness in the representation of the larger land Mammalia is amply compensated in the numbers of heavy aquatic Mammalia, the whales, walruses, seals, and porpoises of the Arctic seas.

During the last half-century, since the geographical distribution of animals and plants has become a subject of more careful investigation among naturalists, these broad zones of the earth's surface, with their characteristic populations and vegetation, have been subdivided, according to more limited and special combinations of organic forms, into narrower zoological and botanical areas. The application of these results to marine life is however of much more recent date, and indeed it would seem at first sight, as if the water, from its own nature, could hardly impose a barrier so impassable as the land. The localization of the marine faunae and florae is nevertheless as distinct as that of terrestrial animals and plants, and late investigations have done much to explain the connection of this distribution with physical conditions.

A glance at the coast of our own continent, starting from the high north and making the circuit of its shores, from Baffin's Bay to Behring's Straits, will show us to what a variety of physical influences the animals who live along its shores are subjected. On the shores of Baffin's Bay, especially on the inner coast of Greenland, where the glaciers push their way down to the very brink of the water, and annually launch their southward-bound icebergs, we shall hardly expect to find a very abundant littoral fauna. On its western shore, where the ice does not advance so far, and a greater surface of rock is exposed, the circumstances are more favorable to the development of animal life. Here abound the winged Mollusks (Pteropods), often swept down to the coast of Nova Scotia by the cold current from Baffin's Bay; the "whale feed," as the fishermen call them, because the whales devour them voraciously. Here occur also many compound Mollusks, especially a variety of Ascidians, and the highly colored stocks of Bryozoa. With them is found the Comatula of the northern waters, one of the few modern Crinoids, and beside these a number of Star-fishes, Sea-urchins, and Holothurians, not differing so essentially from those already described as to require special mention.

Along the shore of Labrador and Newfoundland, the coast is wholly rocky, and especially about Newfoundland it is deeply indented with bays. Here there is ample opportunity for the growth of certain kinds of animals in sheltered nooks. The number of species is, however, much greater along the shores of Maine, Nova Scotia, and New Brunswick than in Labrador, owing no doubt to the milder climate. The beautiful shore of Maine, with its countless islands, and broken, picturesque outline, is very rich in species. Parts of this coast are remarkable for a variety of naked Mollusks, as well as for the great numbers of bright-colored Actiniae, and also for the more brilliant kinds of Holothurians, the Cuvieria, and the like. The latter are especially abundant in the Bay of Fundy, and here also occurs the only Northern representative on our coast of the Sea-fans or Gorgoniae, so common on the shores of Florida.

Farther south, from Cape Cod to Cape Hatteras, the character of the coast changes; it becomes more sandy, and though here and there the aspect is varied by a rocky promontory or a stony beach, yet the general character is flat and sandy. With this new character of the shore, the fauna is also greatly modified, and it is worthy of remark, that while thus far the representative species have reflected the character of animals to the north of them, they now begin to represent rather those of the Carolina shores. South of Cape Cod come in a kind of Scallop and Periwinkle, very different from the larger Scallops found on the coast of Maine and the British Provinces; our Sea-urchin is replaced by the Echinocidaris, with its few long spines, and an entirely new set of Crustacea and Worms make their appearance on this more sandy bottom. And here we must not forget that not only is the aspect of the animal life changed, as we pass from a rock-bound to a sandy coast, but that of the vegetation also. The various many-tinted sea-weeds of the rocky shore disappear almost entirely, and their place is but poorly supplied by the long eel-grass, which is almost the only marine plant to be found in such a locality. Beside its more sandy character, the coast from Cape Cod to Cape Hatteras is affected by the large amount of fresh water poured into the sea along its whole line, greatly modifying the character of the shore animals. The Hudson, the Delaware, the Susquehanna, the Potomac, the James, the Roanoke, and the large estuaries connected with some of these rivers, give a very peculiar character to the shore, and bring down, not only a vast supply of fresh water, but also a large quantity of detritus of all sorts from the land. Under these circumstances life would be impossible for many of the animals which live farther north. The only locality on the North Atlantic shore, where the conditions are somewhat similar, is at the mouth of the St. Lawrence, that great drainage-bed through which the Canadian lakes empty their superfluous waters into the Gulf of St. Lawrence.

The whole coast of the Carolinas, from Cape Hatteras to Florida, is a sandy beach; but though in this respect it resembles that immediately to the north of it, it differs greatly in other features. Comparatively little fresh water is poured into the ocean along this shore, and its more southerly range, instead of being protected by sand-spits like Pamlico and Albemarle Sounds, or broken by estuaries and inlets like the coast of Virginia, lies broadly open to the sea. On its extensive beaches we have the large Pholas, burrowing deep below the surface, and the Cerianthus, those long, cylindrical Actiniae, enclosed in sheaths, with their bright crowns of gayly-colored tentacles; the free colonies of Halcyonoids abound also on this coast, and a new set of Sea-urchins (Spatangoids and Clypeastroids) make their appearance.

Farther south, along the Florida coast, a new element comes in, that of the coral reefs, enclosing shallow channels near the shore, and thus providing sheltered harbors on their leeward side, while on their seaward side they slope steeply to the ocean. Beside this, the reef itself affords a home for a great variety of creatures, who bore their way into it and live in its recesses, as some insects live in the bark of trees. Perhaps a more favorable combination of circumstances for the development of marine life does not exist anywhere than about the coral reefs of Florida, and certainly nowhere is there a more rich and varied littoral fauna, especially on their western shore within the Gulf of Mexico. Here swims the Portuguese Man-of-War, borne gayly along on the surface of the water by its brilliant float, here the blue Velella sets its oblique sail to the wind, and hosts of the lighter and more brightly tinted corals fringe the shore with a many-colored shrubbery. In these waters are also found the blue and yellow Angel-fish, the Parrot-fish (Scarus), and the strange Porcupine-fish (Diodon). Vegetable life is comparatively scanty in these tropical waters, where there are scarcely any sea-weeds, except the corallines or limestone Algae of the reefs. The shore of the Gulf of Mexico, as a whole, has much the same character as that of the Carolinas, until we reach the point where the mountains and plateau of Mexico come down to the coast. From this point to the Isthmus of Panama the coast is again rocky.

Crossing the Isthmus and following the Pacific shore of the continent northward, we find a sandy open shore alternating with rocky beaches as far north as Acapulco. Along this coast there is to be found a great variety of corals, especially Sea-fans, growing on the rocks, but no reef. The Pocillopora, an Acalephian coral, the Pacific representative of the Millepore of Florida, is especially abundant. On the peninsula of Lower California we come again upon a rocky coast, with steep bluffs, extending into the sea. Within the Gulf of California are found, on its sandy coast, peculiar kinds of Sea-urchins, Spatangoids, and Clypeastroids, which occur nowhere else on this coast. From Cape St. Lucas up to the Straits of Fuca, with the exception of the large estuary forming the Bay of San Francisco, there are scarcely a couple of harbors of any consequence. The whole shore is most inhospitable, and the violent northwest winds in summer, and the southeast winds in winter, render it still more bleak and difficult of approach. In consequence of these conditions, the fauna is scanty along a great part of the shore; the best spots for collecting are the beaches, near the head of the peninsula, opposite the islands of Santa Barbara and San Diego, and that within the harbor of San Francisco. On the former, large Craw-fishes abound (Palinurus), akin to those of Florida, though specifically different from them. In the latter, the great amount of fresh water prevents the fauna from being exclusively marine; this harbor is, nevertheless, the great centre of the viviparous fishes, and contains also a large variety of peculiarly shaped Sculpins.

Farther north, between the Straits of Fuca and the island of Sitka, the shore resembles that of Maine, with its many islands, bays, and inlets; a succession of long, narrow islands forms a barrier along the coast, enclosing the shore waters, so as almost to make them into an inland sea. But little fresh water empties upon this part of the coast, and here, where the salt water is little modified by any deposit from the land, but where the violence of the ocean is broken by this barrier of islands, there is a full development of marine life. The shores of the Gulf of Georgia, and those of Vancouver's Island, seem to be especially the home of the Star-fishes. The fauna of this locality has been but little investigated, and yet the number of species of Star-fishes known from there is greater than from any other region; many of them are of colossal size, measuring some four feet in diameter. This coast seems also very favorable for the development of Hydroids, in consequence of which its waters swarm with a variety of Jelly-fishes. The Pennatula, that pretty compound Halcyonoid, with its feather-like sprays, is another characteristic type of this fauna. Beyond this, from Sitka to Behring's Straits, the same rocky coast prevails as in Labrador and Greenland. In Behring's Straits we return again to the forests of beautiful compound Mollusks, or rather to a variety of "representative species," resembling the Bryozoa and Ascidians so abundant in Baffin's Bay. The depth of the water, however, is much less here than on the corresponding Atlantic coast, where, south of Greenland, along the shore of Labrador, the water is very deep, while in Behring's Straits the depth is not greater than from one hundred to one hundred and twenty fathoms. The respective faunae of these two shores are also affected by the difference of temperature, the cold current from Baffin's Bay sweeping down upon the coast of Labrador, while, through Behring's Straits, the warm current from the Pacific pours into the Arctic Ocean.

Thus the whole coast of our continent is peopled more or less thickly with animals. But now arises a new set of inquiries; how far into the sea do these animals extend? how wide is their domain? Do they wander at will in the ocean, or are they bound by any law to keep within a certain distance of the shore? These questions would seem to be easily answered, for wherever we go on the surface of the sea, and as far as the eye can penetrate into its depths, we find it full of life; and yet a closer examination shows that all these beings have their appointed boundaries. Along the shores, animal and vegetable life seems to be distributed in certain definite combinations. Those who are familiar with rocky beaches readily recognize the different bands of color produced by the various kinds of sea-weed growing at given distances between high and low-water-mark. First comes the olive green rockweed (the Fucus), and with it are found barnacles and small Crustacea, myriads of which are to be seen hopping about in this rockweed when the tide is out. Below these are the brown crispy Rhodersperms and Melanosperms, and associated with them are Star-fishes, Crabs, and Cockles. Next in order is the Laminarian zone. Here we have the broad fronds of the Laminaria, the "devil's aprons," as the fishermen call them; in this zone is the home of the Sea-urchin, and here will be found also a few small fishes. Lastly we have the Coralline zone, so called on account of the lime deposit in the sea-weeds, giving them the rigidity of corals; among these the Lobsters make their appearance, and here are to be found also numerous clusters of Hydroids, the nurses of the Jelly-fishes.

This distribution is not casual; these belts of animal and vegetable life are sharply defined and so constantly associated, that they must be controlled by the same physical laws. The first important investigations on this subject were made by Oersted, the distinguished Danish naturalist. He undertook a complete topographical survey of the coast near which he lived, carrying his soundings to a depth of some twelve fathoms, and found that both the fauna and flora of the shore were divided, according to the depth of the water, into bands of vegetable and animal life, corresponding very nearly with those given above. His observations were, however, limited, not extending beyond the neighborhood of his home. It is to Edward Forbes, the great English naturalist, whose short life was so rich in results for science, that we owe a more complete and extensive investigation of the whole subject.

Aided by a friend, Captain McAndrew, who placed his yacht at his disposal, he made a series of observations on the British, Scandinavian, and Danish coasts, and explored also with the same object the shores of the Mediterranean. Not content with sounding the present ocean, he sunk his daring plummet in the seas of past geological ages, and by comparing the nature and position of their fossil remains with those of living marine faunae, he measured the depths of the water along their shores. He collected a vast amount of material, and the results of his labors have formed the basis of all subsequent generalizations upon this subject. Nevertheless he arrived at some erroneous conclusions, which, had he lived, he would no doubt have been the first to correct. Dredging from low-water-mark outward, he found that, from the Laminarian and Coralline zone, the animals began gradually to decrease in number, and that, at a depth of two or three hundred fathoms, the dredge always came up nearly empty. He inferred that at a certain depth the weight of water became too great to be endured by animals, and that the ocean beyond this line, like the land beyond the line of perpetual snow, was barren of life. This result seemed the more probable on account of the immense pressure to which animals are subjected, even at a comparatively moderate depth. A column of water thirty-two feet high is equal to one atmosphere in weight; this pressure being increased to the same amount for every thirty-two feet of depth, it follows that a fish one hundred and twenty-eight feet, or some twenty fathoms below the surface, is under the pressure of almost four atmospheres plus that of the air outside. Wherever tides run high, as in the Bay of Fundy, for instance, where an animal is under the pressure of one atmosphere at low tide, and of three atmospheres at high tide, we see that marine animals are uninjured by great changes of pressure. Yet it seems natural to suppose that there is a limit to this power of resistance; and that there must exist barren areas at the bottom of the ocean, as destitute of life as the regions on the earth which are above the line of perpetual snow. No doubt pressure does influence the distribution of life in the ocean; but it would seem, from subsequent observations, that the boundaries assigned by Forbes were far too narrow, and that the structure of many marine animals enables them to live under a weight, the one hundredth part of which would be fatal to any terrestrial animal.

For some years Forbes's theory was very generally accepted, and the results of Darwin's and Dana's investigations, showing that corals could not live beyond a depth of fifteen fathoms, seemed to confirm it. But, quite recently, facts derived from new and unlooked-for sources of information have given a check to this theory. Commerce has come to the aid of science (rewarding her for the gift first received at her hands), and the telegraph cables, alive with the secrets of sea and land, have brought us tidings from the deep. Dr. Wallich, the naturalist who in 1860 accompanied the expedition to explore the bed of the Atlantic, previous to laying the telegraphic cable, first called attention again to this subject. He brought up various animals, highly organized, from a depth of about nineteen hundred fathoms.

Yet, in spite of this positive evidence added to the former observations of Ehrenberg, and to those of Sir James Ross, who, in the Antarctic Sea, brought up an Euryale on a sounding-line from a depth of eight hundred to a thousand fathoms, naturalists were slow to believe that the distribution of animal life in the ocean was not limited to the shallow depths assigned by Edward Forbes. In the Mediterranean and in the Red Sea, from depths of eighteen hundred to two thousand fathoms, living animals have been brought up on the telegraph wires, not of doubtful infusorial character, hovering on the border-land between animal and vegetable life, but of considerable size, as, for instance, one or two kinds of Crustacea, Cockles, stocks of Bryozoa and tubes of Annelids. When the cable between France and Algiers was taken up from a depth of eighteen hundred fathoms, there came with it an Oyster, Cockle-shells, Annelid tubes, Bryozoa and Sea-fans. As these animals were growing upon it, there could be no doubt that they had their normal life and development at this depth, and since they are carnivorous, they tell also of the existence of other animals with them on which they feed.

The dredge, which thus far has played an important part in zoological researches, is destined to revolutionize many of our accepted theories, if we can judge of its future by the brilliant results of the last few years.

From 1861 to the present time the Swedish government has sent several expeditions to Spitzbergen and Greenland. They carried on dredging operations most successfully to a depth of twenty-six hundred fathoms. For some years past Loven, Koren, and Danielssen, the elder and younger Sars, and other Scandinavian naturalists, have made systematic dredgings along the coast of Norway, which, though not extending below four hundred fathoms or thereabouts, have yet furnished most astounding results.

The United States Coast Survey has, in connection with an exploration of the Gulf Stream, been the first to establish a systematic series of dredgings at great depths, continued during several years. The results have proved conclusively that there exists everywhere, in the deep sea, modified, of course, according to the nature of the bottom and the temperature, a most varied fauna, totally distinct from that characteristic of the shores and of shallower waters. Since 1867 Count Pourtales has had charge of these investigations, first established many years ago by Professor Bache and continued by his successor Professor Peirce. He has dredged across the Gulf Stream between Florida and Cuba to a depth of about seven hundred fathoms, collecting an immense number of marine animals entirely unknown before, and characteristic of the different belts of depth, having a most extraordinary geographical distribution, many of the species being found in Florida, the Azores, the Faroe Islands, and the west coast of Norway.

The English Admiralty has for two summers detailed a vessel admirably fitted for such purposes, intrusting the scientific direction of the expedition to Dr. Carpenter, Professor Thomson, and Mr. Jeffreys. Their dredgings, carried on to the enormous depth of two thousand four hundred and thirty-five fathoms, have in every respect corroborated the conclusions drawn from the collections made by Count Pourtales and the Scandinavian naturalists, who, not content with so thoroughly exploring their own coast, have even sent a ship of war to dredge across the whole Atlantic.

These discoveries only show how much yet remains to be done before we shall fully understand the laws of marine life. But we already have ample evidence that the same beneficent order controls the distribution of animals in the ocean as on land, appointing to all its inhabitants their fitting home in the dim waste of waters.

* * * * *